LENSES ON LEARNI

Supervision: Focusing on Mathematical Thinking

Facilitator Book

Catherine Miles Grant, Barbara Scott Nelson, Amy Shulman Weinberg, Annette Sassi, Ellen Davidson, Sheila Gay Buzzee Holland

Center for the Development of Teaching
Education Development Center
Newton, Massachusetts

DALE SEYMOUR PUBLICATIONS
Pearson Learning Group

This work was supported by the National Science Foundation under Grant No. ESI-9731242 and by The Pew Charitable Trusts. Any opinions, findings, conclusions, or recommendations expressed here are those of the authors and do not necessarily reflect the views of these organizations.

Art and Design: Jim O'Shea

Editorial: Stephanie P. Cahill

Marketing: Maureen Christensen and Douglas Falk

Production and Manufacturing: Mark Cirillo, Alia Lesser

Publishing Operations: Carolyn Coyle, Richetta Lobban

ISBN 0-7652-7028-5

Printed in the United States of America

1 2 3 4 5 6 7 8 9 10 09 08 07 06 05

1-800-321-3106
www.pearsonlearning.com

Lenses on Learning: Supervision: Focusing on Mathematical Thinking

Project Staff, Education Development Center

Catherine Miles Grant

Barbara Scott Nelson

Amy Shulman Weinberg

Annette Sassi

Ellen Davidson

Sheila Gay Buzzee Holland

Glenn Natali

Pilot-Test Facilitators and Sites

Jeffrey Benson
Principal and Director of Educational Services,
Germaine Lawrence School
Arlington, MA

Joanne Gurry
Assistant Superintendent,
Arlington Public Schools
Arlington, MA

Joseph Petner
Principal, Haggerty School
Cambridge, MA

Debra Shein-Gerson
Elementary Mathematics Curriculum Coordinator
Brookline, MA

These pilot tests took place under the auspices of the Education Collaborative of Greater Boston (EDCO), Lesley College, and the Merrimack Education Center

Field Test Sites

Houston Independent School District
Houston, TX

Clark County School District
Las Vegas, NV

Portland Public Schools
Portland, OR

New York Public Schools Community District 2
New York, NY

The University of Wisconsin at Milwaukee
Milwaukee, WI

Project Evaluators, Education Development Center

Barbara Miller

Michael Foster

Project Advisors

Diane Briars
Pittsburgh Public Schools

Nancy Dickerson
Boston Public Schools

Judy Mumme
Mathematics Renaissance K–12

Joseph Murphy
The Ohio State University

Mildred Collins Pierce
Harvard Graduate School of Education

Susan Jo Russell
TERC

James Spillane
Northwestern University

Virginia Stimpson
University of Washington

Philip Wagreich and Kathy Kelso
University of Illinois at Chicago

CONTENTS

Introduction to the Course

Standards-based elementary mathematics classrooms, with their emphasis on mathematical thinking and reasoning, pose new challenges for those who observe in classrooms with the purpose of either assisting or assessing the teacher. The teaching in these classrooms is based on fundamentally different assumptions about what knowledge is and what teaching and learning entail. According to one perspective, learning is the absorption of knowledge dispensed by the teacher or a text. In such classrooms, observers look for evidence of dispensing and absorption in the classroom structures and teacher and student behaviors (for example, clear explanation of new ideas by the teacher, opportunities for students to apply new ideas or practice new skills, students observably "on task" throughout the lesson). In a standards-based classroom, mathematical thinking itself is the focus of classroom activity. In order to understand what is happening there, observers must look for the central intellectual ideas of the lesson and how classroom structures and practices provide opportunities for students to develop those mathematical ideas. Observers' purposes in classroom observations are to assess the quality of mathematics teaching and learning and to support the teacher's continued learning. (Of course, at this time of transition many classrooms are mixtures of both modes of instruction and observers will need to be able to identify and understand both.)

This course is designed to help administrators develop the skills to be effective observers in a standards-based mathematics classroom. It has been developed for use with a wide range of administrators such as principals, mathematics supervisors or coordinators, staff developers, math specialists, assistant superintendents, Title I coordinators, and special needs directors. Through the course, administrators learn to focus on the mathematical essence of a lesson and engage teachers in productive conversations after observing in their classrooms. Administrators also reflect on what *they* can gain from observing in standards-based classrooms and how this knowledge can inform their work more broadly.

This introduction begins with a description of the conceptual design of the course and the pedagogical principles underlying it. Following that is a brief description of each of the eight sessions and a discussion of the materials provided for the course. The final section describes the types of activities found throughout the course and lays out a number of pedagogical considerations that you, as facilitator, will want to take into consideration as you prepare to facilitate the course.

Conceptual Design of the Course

This course is built on the premise that administrators, like teachers, learn new ideas most effectively by actively engaging with intellectually interesting and relevant issues. In the course, administrators confront and resolve dissonances they may experience between their initial understanding of a given classroom situation and new evidence that challenges that understanding. Through such a process, they are effectively thinking in order to learn—constructing their own knowledge of mathematics instruction. Participating administrators will have frequent opportunities to think through a number of issues relating to observation and teacher supervision, providing a rich terrain for continued learning.

Thematic Strands

The course is designed around two thematic strands that develop in interaction through the eight sessions. A third important idea, that of distributed supervisory practices, is considered in the final session of the course.

Strand 1. Developing an Eye for Mathematics Classrooms

In this strand, three aspects of classroom observations are examined: the mathematics content of the lesson, the teaching and learning that are taking place, and the nature of the intellectual community in the classroom.

Developing the capacity to discern the features of a classroom that are central for student learning is not an easy task. Observers must learn to see beyond the physical and behavioral aspects of the classroom in order to grasp the intellectual activity underlying the classroom practices. To support this learning, the course includes an observation tool called the *Observation Guide,* described more fully in the **Pedagogical Considerations** section. Through a systematic set of questions, the Observation Guide helps users focus attention on the mathematics, learning & pedagogy, and intellectual community in a classroom. Participants will practice using it as they view videotaped clips of mathematics lessons and observe actual mathematics lessons in their own schools or districts.

The videotaped clips of mathematics lessons allow for a close viewing of classroom action. Events in a videotaped lesson can be viewed multiple times so that participants can revisit these events in ways they cannot in a real time observation. Further, the very concreteness of the images on videotape makes it possible for both participants and facilitators to point to incidents on the tape and interpret, or reinterpret, them from different viewpoints. The clips also provide grounded images of what teaching and learning in a standards-based classroom may actually look like in practice.

Participants will also use the Observation Guide as they carry out a series of classroom observations and teacher conferences in their schools. These "real time" observations give participants an opportunity to practice using the new observation principles in their own schools.

Strand 2. Rethinking Administrators' Talk with Teachers about Mathematics, Learning, and Teaching.

The course is grounded in the belief that, just as students actively construct new knowledge, teachers must be active participants in constructing their own new knowledge about mathematics, learning, and teaching. While this can happen in professional development programs, it can also happen even more powerfully, as teachers reflect on events in their own classroom. When teachers reflect on the mathematical thinking of children in their classrooms, they become generative learners. That is, their knowledge continues to grow and contributes to an ever-more developed understanding of effective mathematics instruction. The second strand of the course has participants reconsider the pre- and post-observation discussions that they often have with teachers.

In these conferences, administrators can support teachers in developing an orientation of curiosity about their students' mathematical thinking. One way to do this, which we call "collaborative inquiry," has both the teacher and administrator share their curiosity and questions about student thinking. These questions can prompt informal investigations from which both teacher and administrator can construct new knowledge about mathematics, learning, and teaching. In this way, collaborative inquiry extends generative learning to the administrator as well as the teacher. Through the various tasks in the course—working with key mathematical ideas in the elementary classroom, seeing how students make sense of these ideas, and learning to use pre- and post-observation conferences for collaborative inquiry—participants can begin to see how supervisory practices can support generative learning.

Distributed Supervisory Practices

The course focuses largely on the teacher-supervisor dyad, which is likely to be a familiar model for most administrators. In the final session, participants step outside of this model to consider a redistribution of the responsibility and authority for classroom observation across a variety of roles in the school. They explore ways in which classroom observation and teacher supervision can become a context for school-wide professional development for both teachers and administrators.

Mathematics

While this is not a mathematics course for school administrators, mathematics is fundamental to its design. Two central tenets of the course are that mathematics is about a set of *ideas* as well as procedures rather than *solely about procedures* and that observers should focus on these ideas when conducting classroom observations. Thus, participants need to have experience working with essential mathematics concepts from the K-8 curriculum. The mathematics topics we have chosen for this curriculum—patterns and fractions—are key topics in the mathematics curriculum throughout these grades. Pattern work, especially common in the early grades, helps to lay the groundwork for more advanced mathematics such as algebra. Fractions, while an important topic in their own right, are integrally related to work in other areas such decimals, percents, and ratios. We have presented activities for these topics that we hope participants will find relevant as well as interesting and challenging.

In Session 3, participants are asked to consider in some depth the issue of early algebra and algebraic thinking. They are (re-)introduced to the concept of "early algebra" so that they can knowledgably participate in policy and pedagogical discussions on the issue. Familiarity with the idea of "early algebra" is important for several reasons:

- Algebra serves as a gateway for challenging and interesting courses in high school math and science. Students who take these courses have access to more and better academic and occupational opportunities. Recent legislation (i.e., NCLB Act of 2000) has re-emphasized the importance of providing these opportunities to *all* students and has led to efforts to prepare all students to take algebra. The current thinking on what this means for elementary students builds on earlier work of, among others, Project SEED, begun in the 1960's and the Algebra Project begun in the 1980's. Based on the belief that given the proper contexts, virtually all students can learn algebra, these projects were among the first to design mathematics programs that made algebra accessible to all.
- Making algebra accessible for all students means providing elementary students with a good preparation for algebra. Administrators at this level need to have a good grasp of what this preparation entails. They need to realize that the kinds of mathematical inquiry and student sense-making supported by standards-based classrooms help lay the groundwork for supporting students' understanding of the algebra they will encounter later on.
- When administrators have an understanding of what "early algebra" means, they can better support teachers to take a sense-making approach when teaching their students.

In the **Pedagogical Considerations** section, the principles that guide the design of the mathematics activities are described in more detail.

Overview of the Sessions

Each session is designed to be three hours long. It is important to make use of the full three hours as the activities within each session are tightly integrated, and the ideas that underlie them build systematically throughout the course. Thus it is difficult to either skip around among activities or to skip activities entirely. Further, activities as described here take the full three hours. It will be important for participants in your course to arrive early for any socialization they wish to do, so that the course can begin on time.

- **Session 1: Changing Mathematics Classes, Changing Supervision.** Participants are introduced to the thematic strands of the course and the Overall Observation Guide that will serve as a tool in their classroom observations. They discuss the implications for classroom observation and teacher supervision that changes in our understanding of mathematics, teaching, and learning can have.
- **Session 2: Building Intellectual Community in Mathematics Classrooms.** Drawing on the Observation Guide, participants consider what makes for an intellectual community in a mathematics classroom.
- **Session 3: Linking Intellectual Community with Mathematical Inquiry.** Participants continue using the Observation Guide to examine the nature of intellectual communities. The session lays the groundwork for attending to mathematical content by giving participants the opportunity to do an extended mathematics exploration and think about the "early algebra" component of it. Participants begin to think about the second thematic strand, "Talking with Teachers," by considering what they would focus on in their discussions with teacher.
- **Session 4: Observing for Content: Listening to Children's Ideas about Fractions.** Participants begin listening and observing for the mathematical ideas of a lesson and the ways students make sense of these ideas. Participants also consider what is involved for teachers as they make sense of students' mathematical ideas.
- **Session 5: Supporting Generative Learning: How We Talk with Teachers.** Participants continue to observe for mathematical content. They also examine *how* they talk with teachers by analyzing one of their own post-observation conferences. Participants are introduced to the notion of "generative learning."
- **Session 6: Observing How Knowledge is Constructed in Mathematics Classrooms.** Participants shift to looking at the interaction of a teacher's pedagogical moves and student learning. They also view a post-observation conference with the teacher depicted in the videotaped classroom episode to explore more fully the notion of "collaborative inquiry."
- **Session 7: Building Mathematics Understanding: More than the Sharing of Ideas.** Participants continue to explore the relationship between teaching and learning. They also explore further the nature of collaborative inquiry in post-observation conferences.
- **Session 8: Bringing It All Together: Fostering Generative Learning and Distributed Supervisory Practices.** Participants consider their roles as classroom observers and supervisors from a broader perspective. They examine the kinds of authority they bring and look at how the responsibility for classroom observation and supervision can be distributed across roles in schools. In this context, they consider how classroom observation and teacher supervision itself can be a valuable form of professional development that can support generative learning for teachers throughout the school as well as for themselves.

The chart below displays how the two central strands are woven through the eight sessions of the course.

	Strand 1 Developing an Eye for Mathematics Classrooms	**Strand 2** Rethinking administrators' talk with teachers about mathematics, learning, and teaching	Distributing the authority for and responsibility of Supervisory Practice and Supporting generative learning:
Session 1	Introduction to Observation Guide	Changing perspectives on supervision	
Session 2	Focus on Intellectual Community		
Session 3	Focus on Intellectual Community	What to ask the teacher about	
Session 4	Focus on Mathematics Content	What to ask the teacher about	
Session 5	Focus on Mathematics Content	Supporting generative learning: how we talk with teachers	
Session 6	Focus on Learning and Pedagogy	Supporting generative learning: how we talk with teachers	
Session 7	Focus on Learning and Pedagogy	Supporting generative learning: how we talk with teachers	
Session 8			Communities of Inquiry Supervision as a context for school-wide professional development

Organization of the Materials for the Course

Facilitator's Book

The Facilitator's Book contains all of the materials needed to teach the eight three-hour sessions. Each session in the Book consists of the following:

Introduction The Introduction describes the session in general, provides the rationale for the activities in the session, and explains its niche in the larger context of the course.

Overview This short section describes the different activities of the session, the materials needed, and the advance preparation you will need to do. The preparations described include administrative, logistical, as well as intellectual preparation. The intellectual preparation may involve doing any mathematics prior to the session, previewing the video, reflecting on the discussion questions, and reading the **Facilitator Notes** and any assigned articles for each activity.

Activities A "step-by-step" description of each of the activities of the session is provided. This description includes materials needed for the activity, recommended time for each part of the activity, recommended discussion questions, and pedagogical suggestions, such as grouping of participants during the activity.

In addition, "intent" boxes are occasionally placed on the right of the text. These are meant to explain more fully the intent of particular activities or assignments so that you can tailor the activity to the particular needs and interests of the group without losing the principle purpose of the activity.

Facilitator Notes Facilitator Notes accompany each of the activities of the session. They explain the ideas behind the activity and propose responses for discussion questions. The Notes often highlight key ideas that participants should consider as well as some thoughts that participants might share in the discussion. They also provide a synopsis of an assigned article or videotaped classroom episode. In addition, the Notes include an explanation of the mathematics for any mathematical exploration done.

Homework The homework assignment and its purpose are described.

Bridging to Practice Each session ends by having participants engage in a reflective writing exercise. A framework is provided to guide the writing.

At the end of each session are found all of the handouts and homework assignments used in the session.

Readings Each participant receives a book of *Readings,* which contains all of the homework assignments and readings for the course.

Pedagogical Considerations

This course functions as a reflective and intellectual community. What follows is intended to provide some guidelines for facilitating the course as such a community. It describes the pedagogical principles underlying each type of activity (e.g. doing mathematics together, observing videos, etc.) and offers some general pedagogical guidelines for facilitating the activities. Because these descriptions are not situated in the context of specific activities, they may seem abstract upon first reading. In the **Facilitator Notes** that accompany each activity, we often recommend that you refer back to this section as you engage in the different tasks of the course.

Activities

Participants engage in several different types of activities, each of which is described in detail. These include:

- Doing mathematics together
- Viewing and analyzing videotaped clips of teachers and students at work in the elementary classroom
- Using the Observation Guide
- Conducting real-time observations of teachers teaching math
- Participating in reflective discussions of the mathematics, videotaped clips, readings, and classroom observations
- Engaging in reflective writing

Doing Mathematics Together

In Sessions 2 through 7, participants will do mathematics together. The mathematics that they do is taken from the videotaped classroom episodes that they see in that session. Sometimes participants do the mathematics problems exactly as they appear on the videotape. On other occasions, they do an extended mathematical investigation that is designed to probe more deeply into the mathematical ideas that they will view in the classroom episode. The pedagogical reasons for having participants do mathematics together include:

- **Preparation for Viewing the Videotaped Classroom Episode.** Participants will be better prepared to view the classroom episode and to make conjectures about the children's understanding of the mathematics in the lesson. By doing the mathematics, participants can explore the concepts and ideas that underlie the problem, the intent of the problem, and various approaches to solving it. Further, doing the mathematics can prepare participants to anticipate what they will see in the classroom episode, partially substituting for a pre-observation conference that is absent in the videotaped clips.
- **Understanding of K-8 Mathematics from a Standards-based Perspective.** While this is not a mathematics course, one goal of the course is to (re)introduce participants to central mathematical *ideas* found in the elementary curriculum. By doing mathematics, participants can appreciate the complexity of the ideas addressed in the curriculum and can begin to view the discipline of mathematics more generally as a set of ideas rather than being solely about *procedures.* This can help them shift their thinking about how to approach different kinds of mathematical investigations.
- **Experience with Standards-based Pedagogy.** While participants have probably read many of the NCTM and state standards documents, they may not have had concrete experiences to link to such terms as "problem-solving" or "mathematical discourse." Further, they may never have thought of themselves as mathematical thinkers. In fact, they may be most familiar with an approach to mathematics instruction that emphasizes memorization of mathematical facts such as the multiplication tables, and procedures or algorithms such as long division. While such facts and procedures are important, focusing solely on them often masks the underlying ideas about numbers and their relations that make the procedures work.

In doing mathematics together, participants will explore numbers and their relations from a standards-based perspective. They will do this in a way that helps them become mathematical thinkers as they puzzle about complex and interesting mathematical ideas. By doing mathematics *together,* participants can listen to other people's mathematical ideas, articulate their own, reflect on the values embedded in these ideas, engage in collaborative problem-solving, and experience what it is like to be part of a group of people working on mathematical ideas together. Doing mathematics together allows participants to experience mathematics and mathematics learning differently and contributes to creating a reflective community of learners.

It is important for participants to have these experiences if they are to understand the goals of standards-based mathematics instruction, including the mathematical understanding that students are to develop and the teaching approaches that help develop these understandings.

Viewing Videotaped Clips of Teachers and Students at Work

Setting a Climate for Viewing Videotapes of Teachers and Students at Work. In almost every session, you will be showing a videotaped clip of excerpts of a classroom episode in which students talk about their mathematical thinking or they work with teachers in a mathematics classroom. These videotaped clips were chosen for the ideas about mathematics, teaching, and learning they make accessible for consideration and the diversity of teachers and students they depict. Our intent was to present classrooms that look and sound familiar to the participants.

In watching these videotaped classroom episodes, participants will learn to focus on the mathematical thinking that is taking place in the classroom and the instruction supporting the development of students' mathematical ideas. Learning to "see" in new ways requires focusing on the videotape with an orientation of curiosity and discernment, not one of evaluation of the quality of the teaching. Participants should ask themselves, "What can be learned about teaching mathematics from this glimpse of a classroom?

These guidelines will help establish a productive climate for viewing these videotaped clips.

1. Remind viewers that they are seeing only excerpts of a wider learning event. They should resist the temptation to make judgments about the teacher and the classroom based on this limited view.
2. Provide relevant information about the context of the videotaped clip (e.g., age or grade of students, summary of the rest of the lesson from which an excerpt is drawn when available).
3. Be clear about the purpose of viewing the videotaped clip.
4. Where complex mathematics thinking is involved, always have participants work with the mathematics problem themselves first. In this way, participants can focus on the thinking of teachers and students in the lesson rather than understanding the mathematics of it.

Orientation Toward Teachers when Observing. In all of the videotaped clips, participants examine the work of real teachers who are in the process of developing a standards-based approach to mathematics instruction. All of the teachers agreed to have their practices filmed as a contribution to increased understanding of the nature of mathematics instruction. From the videotaped clips available, we chose clips of teachers exploring and wrestling with different pedagogical issues related to the teaching of mathematics because we wanted participants to investigate different aspects of teaching. Analyzing what each of the teachers appears to understand, and what that teacher is still working on or may need to work on is important intellectual work for the administrators in this course. However, when participants discuss teachers' work, it is important that they maintain the same attitude of respect that characterizes the tone of the discussion when considering students' work. Just as students are always working on their mathematical ideas and their energy and creativity is amazing, so are these teachers dedicated to their work and developing their craft as best they can. The purpose of viewing the clips is to develop conjectures about what the teacher understands and knows how to do, and what he or she still needs to learn, not to determine what the teacher is doing, right or wrong.

Ideally, the classroom episodes that participants view would include pre-observation conferences with teachers to provide the information administrators need as they think through what they are viewing. Given this lack of information, participants need to be especially careful to not jump to conclusions about what they observe. For example, while they might think that a particular teacher did not go far enough into the mathematics of an activity, they should keep in mind that they do not know what the teacher's intentions were for that class session and what other factors might have been influential in the teacher's decision-making process.

Showing the Videotaped Clip Twice. The videotaped clips are always viewed two times for several pedagogical reasons. First, participants will be able to notice aspects of the activity and the classroom in the second viewing that they may have missed in the first viewing. This allows for a more in-depth study of the interactions in the classroom. Second, the first viewing allows participants to become familiar with the lesson and the issues that the students and teacher are dealing with so that they can analyze these issues more fully in the second viewing. Thirdly, in the first viewing participants can discuss questions or concerns that may fall outside of the primary focus of the course. For example, participants may have concerns about some classroom management issues or questions about the teacher's use of class time. Having an opportunity to discuss these concerns after the first viewing can help participants be more focused on the mathematical ideas in the lesson in the second viewing.

It is important to understand that viewing the videotaped clip twice is a pedagogical technique designed to provide participants with the practice needed to refine their observation skills. Some participants may find the two viewings repetitive. But, just as a musician repeatedly practices a passage of music or a sports player practices a play many times, valuable learning can come from looking at the same classroom event more than once.

The Observation Guide

We developed an Observation Guide for use while viewing and discussing the videotaped clips of classroom episodes. The Guide is designed to help administrators develop discernment on what to attend to when observing in a standards-based classrooms. It is not intended for use in formal teacher evaluations. The Guide should be used with an orientation of curiosity about what is happening in the mathematics class. When using the Guide in actual classroom observations, participants should use it for the purpose of supporting teachers in their continued professional development.

This Guide provides a structure for looking at the classroom as a whole. It highlights three important and interdependent dimensions of the K-8 mathematics classroom, the mathematics content, learning and pedagogy, and the intellectual community in three horizontal sections of the Guide. These dimensions are presented as analytic categories so that observers can become more discerning in their interpretation of classroom events. A given teacher move or student comment can simultaneously embody all three dimensions. The mirrored columns on the facing pages emphasize the importance of the interplay between the students and teachers. The Guide is meant to help observers become attuned to interdependence of students' actions and teacher's pedagogical decisions. The two columns on each page ask observers to make conjectures, not judgments, about what they notice in the classroom and to offer evidence from which their conjectures arise.

The Observation Guide consists of a list of open-ended questions for each of the three dimensions as they relate to students and the teacher. These questions direct the observer's attention to a particular aspect of the lesson and are specifically designed to highlight important features of a standards-based mathematics classroom.

The Guide is deliberately comprehensive so that it includes a large number of components that would be seen across a range of standards-based classrooms, but not all of which would be found in one classroom. Thus, it is unlikely that any particular classroom would exhibit all of the elements indicated in this Guide at the same time.

What observers see when observing a videotaped classroom episode will be a function of both what is on the videotape and their current understandings about mathematics, teaching, and learning. Part of the purpose of this course is to help observers develop new ideas about mathematics, learning, and teaching so that, over time, they will see more in each videotaped classroom episode and, ultimately, in the classrooms in their own schools.

The Guide functions on two levels:

1. *The Overall Observation Guide.* The Overall Guide will help participants weave all the dimensions of the classroom together as they gain skill at observing the details within each of the categories. Participants will use the Guide in the first session of the course and for the first viewing of most videotaped clips.

Overall Observation Guide					
	Students			Teachers	
Focus Question	**Conjecture**	**Evidence**	**Evidence**	**Conjecture**	**Focus Question**
Math Content					**Knowledge of Content**
What mathematical ideas are embedded in this lesson?					What does the teacher seem to understand about the mathematics?
Learning					**Pedagogy**
Intellectual Community					**Facilitating Intellectual Community**

2. *The Expanded Observation Guides.* In addition to the Overall Observation Guide, there are three Expanded Observation Guides, one for each of the three dimensions (Mathematics Content; Learning & Pedagogy; and Intellectual Community). These Expanded Guides contain the same questions as are found in the Overall Guide, but for each general question, there is a set of sub-questions that prompt the observer to look for more specific evidence. For example, the first question in the mathematics category on the Overall Guide is "What mathematical ideas are embedded in this lesson?" On the Math Content Guide, there are three additional scaffolding questions, "What is the topic?" "What are the ideas within this topic that are being explored?" and "What specific ideas are being explored by different students or groups of students?" These sub-questions help the observer focus on the discrete elements of the classroom episode that can help to answer the general question, "What mathematical ideas are embedded in this lesson?"

Overall Observation Guide					
	Students			Teachers	
Focus Question	**Conjecture**	**Evidence**	**Evidence**	**Conjecture**	**Focus Question**
Math Content					**Knowledge of Content**
What mathematical ideas are embedded in this lesson?					What does the teacher seem to understand about the mathematics?

Math Content Guide					
	Students			Teachers	
Focus Question	**Conjecture**	**Evidence**	**Evidence**	**Conjecture**	**Focus Question**
1. **What mathematical ideas are embedded in this lesson?**					
a. What is the topic?					
b. What are the ideas within the topic that are being explored?					
c. What specific ideas are being explored by different students or groups of students?					

Participants systematically explore each of the three categories in the Guide. In Session 1 they are introduced to the Overall Guide. In Sessions 2 and 3 they focus on the Intellectual Community in the classroom, on the Mathematics Content in Sessions 4 and 5, and on Learning & Pedagogy in Sessions 6 and 7. As they work with each of the Guides, their observation skills will cumulate.

Using the Observation Guide in Practice. The Guide is meant to be used as a set of guidelines for what to attend to in an observation, whether it is a videotaped clip or a real-time lesson. In general, participants' responses to the questions on the Guide will be conjectures or interpretations, for which evidence should be given. Participants should be encouraged to develop their own approach to collecting evidence to support their conjectures. Some participants may like to take notes directly onto a copy of the Observation Guide. Others may prefer to take notes in a notebook and then to use the Guide to organize and interpret their notes.

Take as an example the question, "What mathematical ideas seem confusing to students?" In observing a fractions lesson, an administrator might write, "I noticed that José said that $\frac{2}{6}$ is twice as much as $\frac{1}{3}$ and I suspect that José is confused about how fractions increase." Determining precisely what it is that José understands and what he is confused about would require further investigation. In a real observation, the administrator's interpretation of José's confusion might become part of the post-observation conference.

This focus on the development of conjectures about what is happening in the classroom is part of the overall orientation of this course toward developing "inquiry-based discourse" for supervision. Such an orientation, which encourages teachers to think critically about their own practice, is essential if they are to become generative learners about children's mathematical thinking. It is also a necessary condition if teachers are to develop their own teaching over time. As we noted earlier, administrators can help teachers develop this orientation by having an orientation of curiosity themselves about the mathematical thinking in the classrooms they observe. That is, administrators need to engage in inquiry-based discourse that focuses on the discussion of conjectures and possibilities rather than on definitive answers and scripts for behavior.

Conducting Observations of Teachers at Participants' Schools or School Districts

Participants will be asked to do three classroom observations in their own schools or districts. These observations will include a pre-observation conference with the teacher at which they will gather information about the class and the lesson they are about to observe. Participants will use the Overall Guide as well as the different Expanded Guides during the observations. The expectation is that participants will gain much from using these Guides in real classrooms. Participants will do these real observations between sessions 2 and 3, when they will use the Intellectual Community Observation Guide; between sessions 4 and 5, when they focus on the mathematics content; and between sessions 6 and 7, when they will use the Learning & Pedagogy Observation Guide.

Participants should be told of this requirement in the first session of the class and instructed to choose two teachers to observe. We suggest that they choose teachers who are at different places in the development of their mathematics teaching and are not very traditional or reluctant-to-change. In addition, the teachers selected should be in an off-year in the evaluation cycle so they can view themselves as partners in the administrators' learning.

Using the Pre-observation Conference Questions

We have developed a set of questions for participants to use before classroom observations to help them make sense of what they will be observing. From these questions, an administrator can know what the teacher hopes to accomplish in the lesson, what mathematical ideas will be the focus of the lesson, and which students, if any, have special issues that affect the teaching and learning in the classroom. This information can help administrators bring an understanding to what they observe that would not be possible otherwise. With this increased understanding, an administrator will be able to provide better information about what happened in the class to the teacher in a post-observation conference. Furthermore, these questions can help the teachers themselves be more thoughtful about their goals for a class and their strategies for achieving them. They too can keep track of how well aligned their intended learning goals are with students' actual learning.

The pre-observation questions are:

What topic will you and your students be working on in this lesson?

What do you plan to do in this lesson?

What do you hope to accomplish in this lesson?

What mathematical ideas are embedded in this lesson?

What have you and your students been working on prior to this lesson?

How does this lesson fit into your overall goals for the year?

Are there students who have special issues in the class?

Participating in Reflective Discussions of the Mathematics, Videotaped Clips, Readings, and Classroom Observations

The reflective discussions that are part of almost all of the activities are critical. They provide the opportunity for participants to reflect on the ideas and to come to new and deeper understanding of issues and their own stance toward them.

However, it is not expected that all of the questions in this text will be discussed. They are provided as prompts to initiate a lively discussion. In preparing for each activity, you will want to choose the questions that seem most relevant for your particular group. Being able to do this well requires listening carefully to the ideas that the group raises, paying attention to where each participant is with respect to a set of ideas, and figuring out what seem to be the most fruitful issues to pursue with a particular group.

Supporting Informal Conversations among Colleagues

Some sites choose to offer this course in the morning (e.g., from 7:30 A.M. to 10:30 A.M.) and others in the late afternoon or early evening (e.g., from 5:00 P.M. to 8:00 P.M.). In both of these cases, it can be useful to provide food such as a continental breakfast or light supper. In order to start on time and complete the activities, encourage participants to arrive early for informal conversation. If food is provided, you may want to have it available a half hour before the session begins. This informal social time before the session will give participants a chance to get to know each other and build a sense of community and shared purpose. It can also help to keep participants focused on the substantive work during the session.

Engaging in Reflective Writing

Throughout the course, participants do a range of writing exercises, some of which are done in the session and others are done for homework. Writing is done primarily for the participants' own reflection and learning. It is strongly recommended that participants hand in their writing or copies of it. When it is handed in, the writing can serve several additional purposes.

First, the writing can help you, gain a deeper understanding of where participants are with regards to a set of ideas. This understanding can be useful in planning subsequent sessions. Second, if you, as a facilitator, are able to respond to participants' writing, their thinking can be pushed further. While such responses can be time-consuming, they serve to legitimize the importance of the writing that participants do and encourage continued reflection.

Bridging to Practice

At the end of each session, time is reserved for directed reflective writing, which we call "Bridging to Practice." The prompts are always the same: "Pick an idea that came up today that you think is particularly interesting. What is your current thinking about this idea?" "Where is your school now with regard to this idea?" and, "What are one or two things that you, as instructional leader, will do to move yourself and/or your school along with this idea?"

These questions allow participants to reflect on the session and identify the ideas that they are working on conceptually. They also provide closure to the session, while at the same time suggesting that the work of the session is not finished as participants think about how to continue thinking about, or put into practice, some of the ideas the session has evoked in them.

If participants hand in their original writing, make copies before handing it back. These reflective writing entries can be filed chronologically, either as a group to build a picture of the class or by individual participant to build a picture of that participant's thinking throughout the course. Analyzing the journal entries in this way will help you to think about the ideas that participants may still be working on when they come to the next session. This in turn can help you plan which discussion questions your group might most benefit from discussing.

Homework for Session 1

In preparation for the first session of the course *Lenses on Learning: Classroom Observation and Teacher Supervision,* please do the following:

Read *Changing Perspectives in Curriculum and Instruction,* by Nolan and Francis.

1. Mark those paragraphs or short sections that seem particularly important or thought-provoking, and be prepared to discuss them in class.
2. In addition, choose one of the implications for supervision described on pages 12–18 in the book of *Readings* and quoted below. Drawing on both your own experiences and reflections as a supervisor and/or supervisee, write about why this implication is an important one to consider in the context of reframing the paradigm for teacher supervision.

Implications for supervision articulated by Nolan and Francis:

- Teachers should be viewed as active knowledge constructors of their own knowledge about learning and teaching.
- Supervisors should be viewed as collaborators in creating knowledge about learning and teaching.
- The emphasis on data collection during supervision should change from almost total reliance on paper-and-pencil observation instruments to capture the events of a single period of instruction to the use of a variety of data sources to capture a lesson as it unfolds over several periods of instruction.
- Both general principles and methods of teaching as well as content-specific principles and methods of teaching should be attended to during the supervisory process.
- Supervision should become more group-oriented rather than individually-oriented.

Implication: ____________________

SESSION 1

Changing Mathematics Classes, Changing Supervision

Administrators often face new challenges when they supervise elementary or middle school teachers who are using a standards-based approach to the teaching of mathematics. Just as teachers have to rethink the purposes and processes of their instructional practices when they adapt a standards-based approach, administrators can benefit from rethinking key aspects of their supervisory practice for the standards-based classroom. Two of these aspects are developed in this course.

1) *Developing an eye for mathematics classrooms.* Administrators need to know how to go beyond observing the physical and behavioral elements of the classroom to attending to the mathematical ideas under consideration in the lesson.
2) *Rethinking their talk with teachers.* When teachers focus on making sense of their students' mathematical ideas, they set in motion a process that we refer to as *generative learning,* in which they continually enhance their knowledge about their students' thinking. Administrators can support generative learning through collaborative inquiry, in which they engage in conversation with teachers to actively construct their knowledge about students and classroom practices.

Although teacher supervision is typically seen as a way to help *teachers* learn and to monitor the quality of teaching in the classroom, it can be a medium of learning for administrators as well. The teachers and students who work in the schools provide a living laboratory in which school administrators can continue to broaden and deepen their own understanding of mathematics, teaching, learning, and assessment.

Overview for Session 1

OPENING page 20 20 minutes	**WELCOME AND INTRODUCTIONS** This is a time for welcomes and introductions. This is also an opportunity to let administrators know what they can expect from this session and from the entire course.
ACTIVITY 1 Reflective Writing page 22 20 minutes	**WHAT'S IN IT FOR US?** Participants do some initial reflection on what they gain from conducting classroom observations. Participants write about ways in which classroom observation can support their own practice as they try to bring about changes in mathematics instruction in their schools. They share their writing with a colleague who is sitting close to them.
ACTIVITY 2 Discussion of Reading page 25 40 minutes	**CHANGING PERSPECTIVES IN CURRICULUM AND INSTRUCTION** Participants discuss Nolan and Francis's article, *Changing Perspectives in Curriculum and Instruction* (Reading 1), which they read for homework. The authors examine old and new paradigms of learning and teaching and discuss the implications of these for the work of supervision. They provide a rationale for reframing the paradigm for supervision to reflect changes in classroom practice.
ACTIVITY 3 Video page 29 80 minutes	**DEVELOPING AN EYE** Participants begin the work of reorienting their eyes to what is important in the mathematics classrooms. They are introduced to the Overall Observation Guide, which they use as they view a 15-minute videotaped segment of a fourth-fifth grade classroom, *Fractions with Geoboards.*
CLOSING page 37 20 minutes	**HOMEWORK** Participants read and reflect on Russell et al.'s *Learning Mathematics While Teaching* (Reading 2) in preparation for the next session of the course, whose focus is on Building Intellectual Communities. **BRIDGING TO PRACTICE** Participants finish the session with journal writing that allows them to develop further some of their thinking from the session.

BIG IDEAS

This session explores:

- the nature of observation in a standards-based classroom
- ways of interacting with teachers about what is observed in their classrooms
- the role of classroom observation in supporting instructional leadership

MATERIALS

- ☐ agenda
- ☐ nametags
- ☐ flip charts or overheads
- ☐ markers
- ☐ Handout 1, Homework for Session 1
- ☐ Reading 1, *Changing Perspectives in Curriculum and Instruction* by Nolan and Francis
- ☐ Handout 2, What's In It for Us?
- ☐ Videotaped Clip 1, *Fractions with Geoboards* (DVD #1; Program 15; 17:46–30:37)
- ☐ Handout 3, Overall Observation Guide
- ☐ Handout 4, Homework for Session 2
- ☐ Reading 2, *Learning Mathematics While Teaching* by Russell et al.

PREPARATION

- Well in advance of the first session, send participants the book of *Readings* for the course. Ask them to read *Changing Perspectives in Curriculum and Instruction* (Reading 1) and respond to the questions on page 6 of the book of *Readings.*
- Prepare an agenda for the session on an overhead or on flip chart paper with approximate times for each activity noted.
- Think about the discussion questions for Activity 1. Reflect on your own views of the benefits of classroom observations for administrators who are trying to guide changes in mathematics education in their schools.
- Read Nolan and Francis's article for Activity 2. Think carefully through each of the questions participants have been given. Prepare either a flip chart or an overhead that lists the five implications for supervision that Nolan and Francis discuss in their chapter.
- Become familiar with both the Overall Observation Guide and with Videotaped Clip 1, *Fractions with Geoboards.* This will help you guide the discussion in Activity 3 to further the thinking of the group. Review the description of the Observation Guide provided in the **Introduction.** Review the section *Setting a Climate for Viewing Videotapes of Teachers and Students at Work* (p. 9) in the **Introduction.**
- Write the prompts for the journal writing on an overhead or on flip chart paper. Because the journal writing is repeated in most sessions, you may want to prepare the overhead or flip chart so that it can be used again.

OPENING
20 minutes

Materials
Name tags
Agendas

Welcome and Introductions

Before starting the session, post the agenda where everyone can see it.

15 minutes

Introductions As participants introduce themselves, ask them to say something about the context in which they do classroom observations and about an aspect of supervision that they have been thinking about and would like to explore further.

These topics can highlight that this course is about *ideas in practice* that warrant further examination.

5 minutes

Introducing the Course Discuss meeting schedules; course expectations; participant responsibilities, including homework assignments; and whatever else seems relevant to the circumstances of your group and important to communicate.

This information will provide those in the course with important background information about the participants. That will help everyone put the ideas and reflections shared in context.

Let participants know that they will be required to do classroom observations in their own schools or districts as homework for Sessions 3, 5, and 7. Explain that they should identify two different teachers willing to be observed three times over the span of the course. Ideally the teachers selected will be at different stages in their mathematics instruction, and neither should be a completely traditional teacher or one who is reluctant to consider standards-based instruction. Finally, teachers selected should be in an off-year in the evaluation cycle so that participants can view these observations as collaborative learning experiences, rather than evaluative events.

Present the goals and purpose of the course by explaining that throughout the course, participants will examine two aspects of classroom observation and teacher supervision:

- reorienting the focus of classroom observations to look at the central aspects of mathematics and teaching
- rethinking the pre- and post-observation conferences in order to foster an environment of collaborative inquiry

Explain that in this first session, participants will be introduced to these two strands and to the Observation Guide that they will be using throughout the course.

Mention that although the course focuses largely on the supervisor-teacher relationship, this first session introduces the idea of expanding the pool of educators involved in classroom observations within the school, an idea that will be explored in greater depth in the last session.

Facilitator's Notes
Opening

Overview

As described in the **Introduction,** your job as facilitator is to orchestrate discussions during which participants can consider new ideas about mathematics learning, and teaching that are embedded in standards-based mathematics education. The questions in the **Facilitator Notes** are meant to prompt such discussions.

The **Facilitator Notes** are not intended to be comprehensive; rather, they offer insights into the key ideas of each session and suggestions for discussion questions. As a facilitator, you will want to think through what other ideas might come up, given the makeup of your own particular group. This will be more intuitive as you come to know the participants better.

ACTIVITY 1
Reflective Writing
20 minutes

> ***Materials***
> Handout 2

What's In It for Us?

3 minutes

Introducing the Activity Tell participants that, while we often think about supervision as facilitating teacher growth, they will begin this course by considering and writing about the following two questions:

> *What do you need to know in order to be a good observer?*
>
> *What do you, as an administrator, gain from conducting classroom observations?*

Explain to participants that they have 7 minutes to write and 10 minutes to share their writing and thinking with a colleague.

> Writing allows participants to reflect on ideas and clarify their own beliefs about them. Their writing can give you, the facilitator, an understanding of their current perspectives on supervision.

7 minutes

Individual Writing As participants write, take note of ways in which participants approach this task. Who dives right into the writing? Who needs clarification and what type of clarification? This is part of the process of your coming to know the participants in your course.

> Reflective writing helps participants examine the understandings they bring to supervision and the dimensions of their work that thoughtful supervision can enhance.

10 minutes

Sharing with a Colleague Invite participants to share their writing with the person next to them and to describe briefly any additional thoughts and reflections that come to them. As participants talk with one another, circulate among them, listening to the content of what they are sharing and the tone they use. This will help you become more familiar with the understandings about supervision and its relationship to their work that participants bring and is part of your own ongoing process of making sense of participants' ideas.

Collect these writings, and make copies for yourself. Let participants know that you will return their writing at the the beginning of the next session. These writings can help you get to know the participants and their thinking about these questions more quickly.

Instruct participants to keep this writing in an easily retrievable place in their notebook for later in the course, when they will revisit the question about what they gain from classroom observations.

Mention that in the next activity participants will discuss the article by Nolan and Francis that they read for homework.

Overview

One benefit participants can derive from this course is a strengthening of the foundation for the other work they do in their schools. They will have access to the content of their work as instructional leaders in its most direct form: students' (mathematical) thinking and teachers' learning in the context of classrooms. This core experience is essential if administrators are to be effective instructional leaders.

Sharing Reflective Writing

♦ **What do administrators need to know in order to be good classroom observers?** Much of this course will focus on this question. Participants will use the Observation Guide as an instrument to help them develop and extend their ability to make connections between content and pedagogy in elementary and middle school mathematics classrooms. At this point, writing about this question is primarily intended to 1) help participants begin to focus their thinking on what is important to attend to, and 2) provide you with insights about the perspectives they bring to the course. They will have the opportunity to revisit this question in the final session of the course.

♦ **What can administrators gain from conducting classroom observations?** The work demands of administrators have distanced many of them from the classroom. Many may feel disconnected from the classroom and may have lost the feel of teaching—the joys and possibilities as well as the challenges. Without recent direct classroom experiences, administrators may make decisions or suggestions that are not grounded in the realities of the classroom and are not appropriate or relevant to teachers' situations.

If administrators are to provide intellectual leadership in schools, especially the leadership necessary to bring about substantive changes in mathematics instruction, they need to understand these changes at the classroom level. Given their extensive responsibilities, administrators need to find ways to work efficiently to gain this knowledge. They can use the supervisory process, which is already part of their administrative responsibilities, as a context for deepening their understanding of these changes. In this course, participants will come to view the supervisory process as an avenue through which they can:

- **Deepen their understanding of standards-based mathematic instruction as a foundation for making wise choices.** If administrators want changes to go beyond the purely mechanical implementation of outer features, they need to develop discernment for what is important in the proposed changes. They need to understand the dimensions of teaching mathematics in standard-based classrooms and have an appreciation for the challenges and dilemmas that teachers face in such classrooms. They also need to develop sensitivity to what constitutes effective teaching practices and what might be problematic. For example, administrators should appreciate the dilemmas teachers face in deciding when it is appropriate to step back and make room for student ideas and when it is constructive to step in and guide the discussion towards the exploration of important mathematical ideas. Administrators should also be able to determine what is a valid mathematical argument worth pursuing. In addition, administrators should appreciate that there are rarely clear-cut answers to many of the challenging questions and dilemmas teachers face on a daily basis. Rather, teachers have to be able to make wise decisions and judgments in the moment, and a number of different actions might make sense at any given time.
- **Enhance their credibility as effective leaders.** If they are to lead effectively, administrators need to know intimately the work of the teachers and students in their schools. They need experiences grounded in actual classrooms

to bring this kind of teaching and learning to their teachers. Only then can they engage in productive conversations about implementing the necessary changes for a standards-based classroom. The more closely linked administrators are to the work of the teachers and the students in their schools, the more they gain in both credibility and in their ability to tap into teachers' developing skills and understandings in viable and meaningful ways.

- **Influence practice.** When administrators attend to teachers' and students' engagement with mathematical ideas, they underscore the importance of mathematical *understanding*. By conveying a keen interest in and an accurate knowledge of both the ideas behind mathematics reform and the current realities in their school, administrators can bring professional and moral authority to their efforts to shape the ideas in the school. (These ideas about sources of authority and influence will be explored more extensively in Session 8 of this course.)

A number of participants may have signed up for the course expecting to learn and think about *teacher* learning. They may not have anticipated that a substantial piece of this work involves deepening their own understanding of what is important in mathematics, teaching, and learning. It may take some sensitivity on your part to help them come to terms with the focus on their own learning that is central to this course.

Reflective Writing

Participants will refer to the writing they do here when they do a reflective writing assignment for homework in preparation for the final session. For that assignment, participants will assess their own learning trajectory in the course and consider how their own thinking about classroom observation and supervision may have changed.

ACTIVITY 2
Discussion of Reading
40 minutes

Materials
Handout 1
Flip chart or overhead
Markers

Changing Perspectives in Curriculum and Instruction

10 minutes

Contrasting Old and New Paradigms of Learning and Teaching Review the characteristics of old and new paradigms for learning and teaching as presented by Nolan and Francis. You might find it helpful to tape two pieces of flip chart paper to the wall (or to to divide an overhead into two columns), one labeled "Old Paradigm" and the other labeled "New Paradigm." Ask participants to suggest one or more characteristics, and write them in the appropriate column as they talk.

20 minutes
Whole group

Discussing Implications for Supervision Lead a discussion of the five implications for supervision presented by Nolan and Francis on pages 12–18 in the book of *Readings.* Use the implications listed on the overhead or on flip chart paper as a guide.

For each implication, ask participants questions such as the following:

> *Who chose this implication? Share your thinking about it.*
>
> *Why do you think it is important to keep this in mind when supervising teachers? How does this relate to your work?*

After getting input from all who focused on this idea, ask the group at large about any further thoughts the implication or the discussion raised for them.

Keep moving through the five points in this way until you have addressed all of them. It may be that some of the implications will not have been selected by anyone. If that is the case, discuss first those that participants chose, and then look at the points that no one wrote about. Time permitting, explore any implications that were not selected.

This discussion can help participants make the ideas presented by Nolan and Francis their own. By sharing their own reflections about particular points as they relate to their own experience as supervisors and/or supervisees, they continue the process of making meaning of them for themselves.

10 minutes

Relating Nolan and Francis to the Structure of the Course Remind participants that the two principal strands of the course are linked closely to the following ideas laid out by Nolan and Francis:

- *Developing an eye for what is important in changing mathematics classrooms* reflects the importance of attending to content-specific principles and methods of teaching.
- *Rethinking ways of interacting with teachers about what is observed in their classrooms* draws on the view of teachers as active constructors of knowledge and supervisors as collaborators in creating knowledge.

A related theme, that of expanding the responsibility for supervision to other members of the school community, correlates with Nolan and Francis's emphasis on collegial learning through group-oriented supervision. When supervision becomes more group-oriented, with more members of the school community participating in observations and supervision, there can be a greater emphasis on collegial learning.

Point out to participants that the ways in which they will be exploring an expanded or more distributed version of supervision in Session 8 will go beyond this image of group learning articulated by Nolan and Francis.

Finally, make a bridge to the next activity with a statement such as the following:

> *A more distributed version of supervision makes it possible to draw on the expertise of different members of the school community; still, it is very important for people in formal supervisory roles to have an informed perspective on what is at the core of mathematics learning in the classroom.*

Explain that in the next activity they will begin this exploration by viewing a videotaped classroom episode.

Facilitator's Notes
Activity 2

Overview

This reading introduces participants to some recent ideas about the nature of learning and teaching. These ideas can help redefine what constitutes robust mathematical understanding for students, which in turn dictates what teachers need to know and do to develop students' mathematical thinking. These ideas can also have implications for teacher supervision. That is, not only must teachers make fundamental conceptual changes related to teaching practices, but administrators must also reconsider their conceptions of the nature of teacher supervision. In addition, just as teachers must be sensitive to the uniqueness of each student, administrators must be sensitive to the uniqueness of each teacher. These are some of the important ways in which instructional leadership, as exhibited in the supervisory process, is changing.

Summary of the Reading

Traditional and Emerging Paradigms of Learning and Teaching

Nolan and Francis set their discussion of teacher supervision in the context of Sergiovanni's concept of *mindscapes,* or mental frameworks made up of beliefs and assumptions that influence behavior. They begin by examining traditional views of the learning-teaching process (pp. 8–9) and contrasting those with newer perspectives on teaching and learning (pp. 9–11). In describing these more recent paradigms, they look at new theories of learning; in particular, the cognitive processing of information. They state that "one of the teacher's most important tasks must be to explore the conceptions that learners bring with them to the classroom and help them achieve a new, more refined understanding of those concepts" (p. 9).

This course is designed to give participants extensive experience in observing how teachers do this by observing and discussing a number of videotaped classroom episodes and doing guided observations in their own schools.

Although some participants may have already taken the course *Lenses on Learning: A New Focus on Mathematics and School Leadership* or another conceptually-based professional development course, this may be the first opportunity for many to examine closely the "below-the-surface" changes in standards-based mathematics classrooms. Nolan and Francis provide a framework and a language for thinking about some of these changes. The various activities in this course—analysis of videotaped clips and one clinical interview with a teacher—will encourage participants to explore these changes in the context of actual classroom practice.

It is important to stress with participants that, in contrasting "old" and "new" views of learning and teaching, "old" does not mean "bad," and "new" "good." It might be more helpful to think of the new paradigm as affording greater insight and depth into how learning and teaching actually happen in practice. The old paradigm can still offer useful principles to guide teaching, such as the importance of coherent lessons, of teachers articulating ideas well in order to avoid confusing students unnecessarily, and of constructing accessible representations of complex ideas.

Implications for Supervision

Nolan and Francis list five implications for supervision. As you facilitate the discussion about these implications, you may find the following notes helpful:

1. Teachers should be viewed as active constructors of their own knowledge about learning and teaching. Here the emphasis is on giving teachers the time and opportunity to construct their own "deep, personalized understanding of the model" (p. 13) and to relate concepts and practices to their existing knowledge and experience. Thus, the process of supervision becomes a context for generating knowledge through inquiry and experimentation.

2. Supervisors should be viewed as collaborators in creating knowledge about learning and teaching. Nolan and Francis stress a shift in the view of the supervisor from "a critic" (p. 14) to a more collaborative relationship, in which the supervisor is a co-investigator of learning and teaching with the teacher. Supervisors can be key sources of support for teachers by constructing their own knowledge about learning and teaching. From this stance, the supervisor's classroom data collection can provide the information on which teacher and supervisor can reflect together, which is part of the process of making sense of students' and teachers' work in the classroom. Through collaborative inquiry, teacher and supervisor can go beyond discussing observable behaviors to examining the rich and complex issues that drive the teaching and learning enterprise.

Some administrators may come with a different interpretation of what "collaboration" means and may find it challenging to appreciate fully the sense in which Nolan and Francis use the term. As you facilitate this part of the discussion, pay attention to how participants interpret the meaning of collaboration, and be prepared to push their thinking on how supervisors can be collaborators in creating knowledge about learning and teaching.

3. Data collection during supervision should use a variety of data sources to capture a lesson as it unfolds over several periods of instruction. The classroom data can be used by the teacher and the supervisor to reflect on the learning that is taking place in the classroom. Multiple data sources, including student work, can offer useful insights into the thinking processes of the students in the classroom and into the collaborative work of teachers and students.

4. Both general and content-specific principles and methods of teaching should be attended to during the supervisory process. Here Nolan and Francis reinforce the importance of shifting the focus in supervision away from one that is attuned to general behaviors to one in which reflections on content-specific strategies and methods also play a central role. Throughout the course, participants will have frequent opportunity to attend to the nature of mathematical thinking and understanding being expressed by students and the ways in which teachers interact with students about their understandings.

5. Supervision should become more group-oriented. Here Nolan and Francis point to the benefits that teachers can derive from being engaged in group processes of supervision. Teachers can gain from one another's experiences and insights in much the same way that students gain from well-structured group work with other students. Nolan and Francis suggest that supervision be redesigned as a group process in which supervisors and teachers work together to achieve interdependent group and individual goals and ultimately improve student learning.

Nolan and Francis's notion that supervision can be viewed as an activity to which many people in a school can contribute (through group supervision, peer coaching, or colleague consultation) is one aspect of shared learning across different roles. This idea will be discussed at length in Session 8.

It is important to note that some participants may be very concerned about the challenges that their responsibilities as evaluators place on their efforts to use supervision as a context for both their own and their teachers' professional development. You will want to acknowledge the dilemmas that evaluation poses while encouraging participants to hold this conversation for the last session, by which time they will have developed a range of additional perspectives to bring to the discussion.

ACTIVITY 3
Video
80 minutes

Materials
Videotaped Clip 1
Handout 3

Developing an Eye

5 minutes

Introducing the Videotaped Clip Tell participants that they will view a videotaped segment of a classsroom lesson.

Explain to participants that in viewing and discussing the video segments in this course, they are not to critique the teaching depicted there. Although they might be tempted to do this, stress that they simply will not know enough to make valid judgments about the teaching they observe. Teaching is complex, and from the perspective of an outside observer, there are always other moves that might have been made. An outside observer who sees only a short excerpt of a lesson, such as what participants will see in these video segments, will not know enough about the backgrounds of the students in the class, the context of the lesson, or the teacher's thinking to make valid judgments about whether the teacher's decisions were good ones.

Before viewing the videotaped segment, give participants the following background information:

- This is a 15-minute segment.
- The class is a grade 4–5, team-taught, open classroom in Tucson, AZ.
- The fourth grade teacher, Constance Richardson, is initially shown introducing an investigation to the fourth grade class and then circulating among small groups of students as they explore different ways of making halves using geoboards.
- Geoboards are boards with 25 short pegs arranged in 5 rows of 5 (making it a 4×4 square). By stretching rubber bands over the pegs, students can make different shapes and explore geometric relationships.

Participants will view the video once and then discuss their initial impressions. They will then view it a second time, using the Overall Observation Guide to focus their viewing on particular parts of the lesson.

The purpose of the first viewing of the videotaped clip is twofold.
1) Participants can share their initial impression. Then they will be able to focus more fully on the guided observations done during the second viewing.
2) You can gather important information about what participants notice and find important in classrooms.

You many wish to refer back to the section in the **Introduction** that discusses the overall rationale for viewing videotapes twice.

25 minutes

First Viewing of the Video and Discussion (17:46–30:37) Show the video a first time, and then pause briefly to have participants discuss what they saw and what their impressions were of what they saw. You might ask:

What happened? What did you see and hear?

What were your impressions of what you saw and heard?

Try to keep this initial discussion brief so that you have plenty of time for more substantive processing after the second viewing.

5 minutes

Introducing the Observation Guide Distribute Handout 3, the Overall Observation Guide. Tell participants that this is a tool to support their own learning as they reorient their focus when observing in mathematics classrooms. In this initial experience with the Guide, they will be using the tool in its entirety to get a general sense of what it is designed to help them attend to. In subsequent sessions, they will focus on particular categories of the Guide and will use expanded versions that include questions designed to help them focus on specific aspects of the general questions.

Several points will be helpful in orienting participants to the use of the Guide.

- This is not an evaluation tool. Rather, it is a tool for supporting the professional development of teachers and administrators. This is part of the wider process of rethinking supervision, with a focus on what to look for when we observe in classrooms and how to relate to the teacher about what we see.
- One side of the Observation Guide focuses on students, and the other side focuses on teachers. Often the two sides are related (for example, on the Student side, *What mathematical ideas seem to be confusing to students?,* and on the Teacher side, *What are instances of productive ways in which the teacher addresses students' confusions?*).
- The questions are organized into three categories, which reflect important principles of reform in mathematics:

  Math Content/Teacher's Knowledge of Math Content

  Learning/Pedagogy

  Intellectual Community/Teacher's Facilitating of the Intellectual Community

Emphasize the importance of the pre-observation conference to appreciate fully what is to take place in a lesson. Acknowledge the constraints of the videotaped episodes that do not include pre-observation conferences, thereby limiting what participants can know about the classrooms they view. In order to address this constraint, participants will do the mathematics in the video prior to viewing it so that they have some understanding of what the lesson is about and what mathematical ideas are embedded within it.

20 minutes

Second Viewing of the Video (17:46–30:37) Organize participants into three groups, each of which will focus on one category of the Guide (Math Content/ Knowledge of Math Content; Learning/Pedagogy; or Intellectual Community/ Facilitating Intellectual Community). Note that this grouping will not be leading to small group work; rather, it will ensure that all categories are examined.

Give participants a few minutes to review the questions for their category, and then show them the video again. After viewing the segment a second time, give participants a few minutes to jot down notes about what they saw with reference to the questions in their section of the Observation Guide.

25 minutes

Whole Group Discussion Invite participants to report on what they saw by category. You might begin with a question such as the following:

> *What did people who focused their observation on math content see from both the students' and the teacher's perspective?*

Encourage participants to give specific examples (i.e., data or evidence) to illustrate their comments. Where appropriate, probe for deeper reflections and help participants make connections among ideas and between student and teacher moves.

Continue discussing the participants' observations and comments for the Learning/Pedagogy category and the Intellectual Community category.

You can use the final moments of this discussion to ask participants to comment on their first experience using the Observation Guide. You might ask:

> *What was your reaction to using the Guide?*
>
> *What areas did it help you to attend to that you otherwise might not have focused on as much?*

Summary of the Videotaped Clip

Ms. Richardson's fourth grade class engages in a rich exploration of the concept of one half. The episode begins with Ms. Richardson discussing with students their understanding of *one half.* The verbal descriptions they give indicate that they think of one half as meaning "the same." They volunteer the following phrases: "same size," "same length," "split it between the middle." When she shows the students a square 4 × 4 geoboard split into half by separating out eight smaller squares in the middle, they determine that that isn't half. (It would have been interesting to return to that question at the end of the class session, after students had had the opportunity to explore many different versions of half with the geoboard.)

Then, when Ms. Richardson follows her students' suggestion to divide the geoboard in half by putting a rubber band down the middle of it, she asks them, "Is this half? How do we know that? How could we prove that without counting in the middle?" Together, the students and the teacher come up with the idea of using the 16 small squares on the geoboard as a way to compare the halves. They determine that each half should have 8 small squares.

Next, Ms. Richardson asks whether there is another way they might show one half. A student suggests drawing a diagonal line across the geoboard. Another student points out that that would give a combination of squares and triangles to compare. They combine the 6 squares and the 4 triangles to show that there is still the equivalent of 8 small squares in each half of the geoboard when it is divided diagonally.

Ms. Richardson then sends students off to work on their own investigations using geoboards. Their task is to figure out different ways of making halves of the geoboard. The camera follows Ms. Richardson as she interacts with students who are clustered at their tables. It focuses on a fairly long exchange between Ms. Richardson and a group of students exploring the question of whether one can prove that a shape is half of a geoboard by counting the number of pegs on each side. The interesting mathematics they get into ultimately comes into their proof that the following geoboard is cut in half:

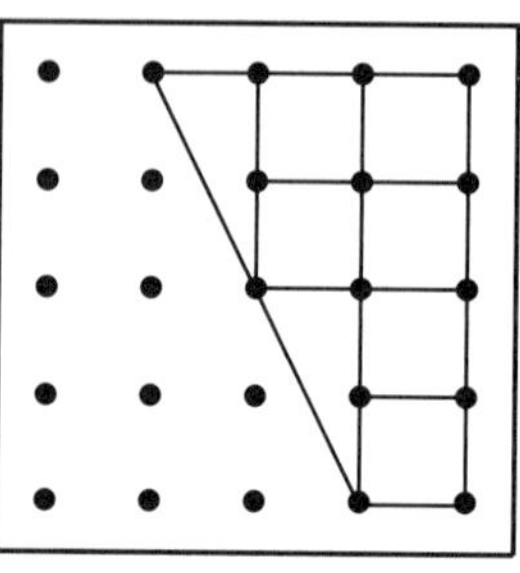

Figure 1

They build up to this by exploring the equivalency between half a rectangle cut on the diagonal (Figure 2) and half a rectangle cut vertically into two squares.

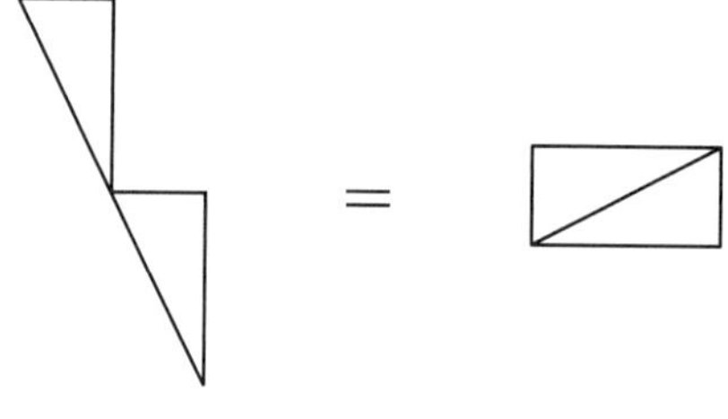

Figure 2

Finally, the class gathers back together and students share the variations on half a rectangle that they have come up with (Figures 3, 4, 5, 6, 7).

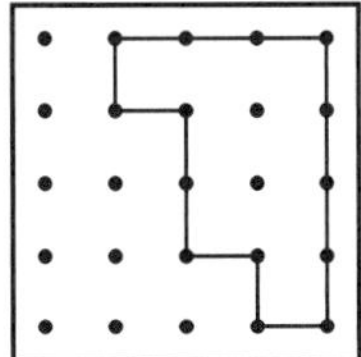
Figure 3

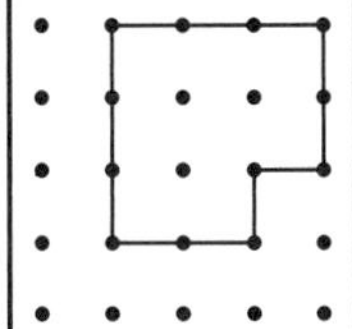
Figure 4

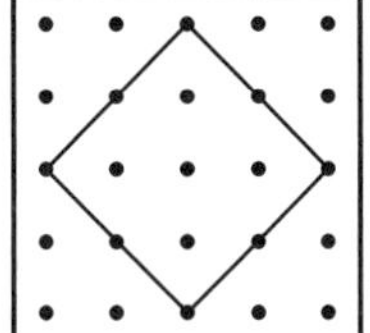
Figure 5

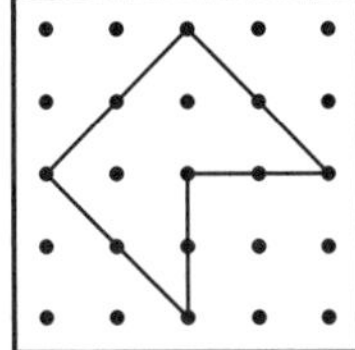
Figure 6

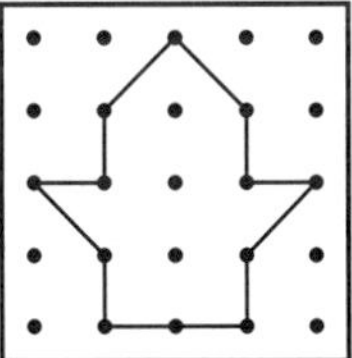
Figure 7

The videotaped episode ends with Ms. Richardson assigning her students an entry in their math journals.

Discussion of the Classroom Episode After the first viewing be sure to ask for specifics from the clip that illustrate the points that participants make. Consider their observations to be conjectures for which they need to cite the evidence. This helps set a tone of genuine inquiry among participants by grounding the discussion in observable features.

It will be very instructive for you as facilitator to take note of what participants say after their first viewing of the videotape. To what extent are their comments or conjectures general impressions about class structure and teacher delivery, and how much do they attend to the mathematical thinking that is taking place? What opportunities for deepening students' understandings of fractions do they notice? It should be interesting to track the evolution of participants' comments about the videotaped clips they view as the course progresses.

Using the Observation Guide for the Second Viewing

Because this is the first time that participants will use the Overall Observation Guide, it is important for you to pay particular attention to the use they make of it. Many participants may be tempted to view the Guide as an evaluation tool rather than a learning tool. You may need to remind participants regularly that the Guide is designed to help them refocus their attention on learning and teaching issues as well as to the content of the lesson.

♦ **Math Content Explored by Students/Teachers' Knowledge of Math Content** This section of the Observation Guide draws attention to the actual content of the lesson: the mathematics in the lesson as well as the teacher's understanding of that content. In the viewed lesson, students work on the idea of "fair shares" in fractional halves in the context of an area model for fractions (where fractions that are less than one are described as portions of a square). They are working on partitioning an area into fractional parts by dividing the area equally. An important concept is that, although two halves of a square are not necessarily congruent (and may not look alike), they will still be equal in area and will represent equal fractions. The many examples for dividing the geoboard into halves that students come up with demonstrate this concept.

This concept of *constancy of halves* is also explored in the small group exploration. The students in the group cut a two-square rectangle diagonally and then explore and ultimately prove that one triangle is equal to a square because each is half of the rectangle (Figures 1 and 2). In addition, this small group exploration helps students realize another central idea in fractions—that a fraction can represent different amounts when the referent wholes are different sizes. In this case, students were exploring *equivalent halves* of the rectangle, which itself was *half* of the original square formed on the geoboard.

The students whose work is displayed in Figures 5, 6, and 7 display a systematic approach to their series of geoboard halves. They begin with a diamond (which the student calls a triangle when she shares) and then explore how extending the diamond on one side and reducing it on the other still leaves them with equivalent halves.

In a follow-up interview after the class session, Ms. Richardson says, "We were trying to get them [the students] to understand what half is without its being a picture that's the same. When we first started talking about halves, they said 'it's the same size, equal length.' We wanted to have them know that half doesn't always have to look the same." This long-term agenda that Ms. Richardson articulates is an important aspect of work with fractions—helping students get past the frequently encountered examples of common fractions in order to become clear about the features of fractions that really are central: in this case, that the parts are divided into equal-sized (but not necessarily equal-shaped) parts.

Participants may wonder whether this exploration of halves is appropriate for students in the fourth grade. Participants can note that in this investigation of halves with geoboards, these students continue to develop ideas central to understanding fractions. It would be interesting, though, to have a conversation with Ms. Richardson about what future work with fractions she intends to undertake with her students. For instance, this activity can serve as a good foundation for exploring further the following issues with an area model of fractional relationships:

- The relationships among fractions with the same numerator and different denominators (e.g., $\frac{1}{2}, \frac{1}{4}, \frac{1}{8}, \frac{1}{3}, \frac{1}{6}$)
- The comparative sizes of the missing piece of fraction (e.g., $\frac{2}{3}, \frac{3}{4}, \frac{4}{5}, \frac{5}{6}, \frac{6}{7}, \frac{7}{8}$)
- Fractions in which the numerator is larger than the denominator

In talking with Ms. Richardson, it would also be interesting to find out what other models she might use for exploring these different fractional relationships (e.g., fractions as divisions of quantities).

♦ **Learning/Pedagogy** This section of the Observation Guide encourages participants to comment on the effectiveness of the teacher's pedagogical moves. Prior to doing so, you should review the *Orientation Toward Teachers when Observing* section of the **Introduction,** in which the attitude of the observer toward the teacher is discussed.

In the reflections she makes after class, Ms. Richardson comments on the development of her students' thinking about halves as the lesson progresses. She notes her students' initial understanding of fractions as implying "equality" or "sameness" and notes their increasing differentiation between *equality of area* and *equality,* or *congruence, of shape.*

Ms. Richardson encourages individuals and small groups to test their hypotheses. She asks, "Can you always use pegs to prove half? How do you know that? Build another half, and see if you can use half to prove it." While investigating the hypothesis that two halves of a square will always have the same number of pegs (a student approach that does not appear to draw on the concept of area as a reliable measure), a small group of students begins to explore the idea of *constancy of halves* (in this case, halves of a two-square rectangle, each half having different shapes but the same area). Ms. Richardson helps her students develop their reasoning skills and multiple strategies to construct proofs of equivalence of area.

Participants might comment that Ms. Richardson could have highlighted the systematic approach to developing the series of halves in Figures 5, 6, and 7. Some may note the important mathematical thinking at play in each group's solutions. Others might point to the complexity of thought suggested in one group's thinking compared with that of another group, such as the group that produced Figure 4, in which the halves seemed to be generated by counting numbers of squares. You might want to encourage participants to turn their observations into conjectures about the teacher's knowledge of mathematics and/or teaching and learning.

Participants might also note that it would have been interesting for the teacher to revisit at the end of the class the initial example of half (8 squares outlined in the middle of the geoboard, with two vertical columns of four squares on either side as shown in Figure 8). What might students say about the example that they had been so sure was not a half after their explorations?

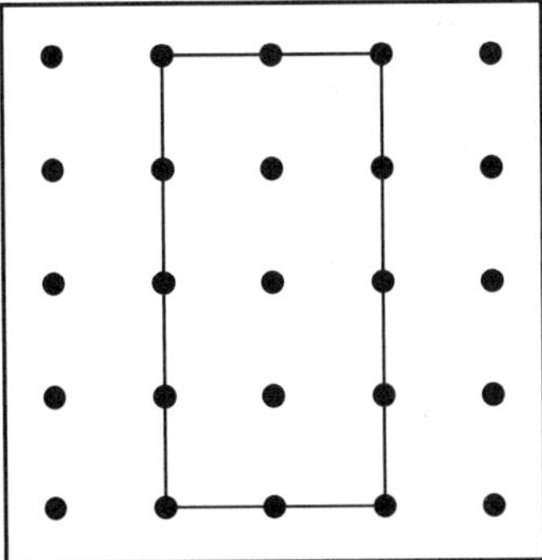

Figure 8

Participants might mention that Ms. Richardson could have used the vocabulary of area more effectively in order to help students see that they were using this measure as a basis for comparison.

Some participants might also be concerned that Ms. Richardson did not correct students' misapplied vocabulary: one used "triangle" instead of "diamond," and another used "square" instead of "rectangle." Here, you could help participants see their concerns as conjectures about Ms. Richardson's content knowledge or pedagogical purpose. Ask participants to think about how they might find out from Ms. Richardson what her reasoning was for these choices. For example, she might have felt that correcting misused terminology *at that time* would shut down the students' thinking and interrupt the flow of the discussion. Teachers are faced with making these kinds of delicate choices every day.

Some participants in your group may note that although this is a racially mixed class, the majority of students Ms. Richardson calls on are Caucasian. This is especially true during the initial class discussion about how to represent one half on a geoboard. If this issue is raised, it would be important for you to remind your group that this classroom episode has been edited, like all others in this course. Their concerns are valid, but we cannot assume that the students of color do not have opportunities to explain their thinking because we see only a small part of the class. Nor can we second-guess the decisions that Ms. Richardson made about whom to call on in this class discussion. Because teachers take into consideration many variables at any given moment, it would not be fair to assume, on the basis of this footage, that Ms. Richardson was providing more intellectual opportunities for the Caucasian students in her class. Later in the footage, when groups present their ways of dividing the geoboard in half, students of color are featured.

If this issue is raised in your group, it would be important to facilitate a discussion of it. You may wish to invite participants to think about what related questions they might ask Ms. Richardson if they were her supervisor. Possible questions include how she groups her students or how she decides on whom to call. Participants might also want to ask about Ramone's role as "official prover." Although some participants might point to this as an example of Ms. Richardson attending to a range of students, even those who aren't directly sharing, other participants may want to know more about Ms. Richardson's thinking about Ramone.

♦ **Intellectual Community/Facilitation of Intellectual Community** This portion of the Observation Guide asks participants to look at the nature of the intellectual community in the class and the ways the teacher supports it. In this videotaped clip, there is evidence of students' deep engagement with mathematical ideas. Students generate ideas about how to prove that a portion of the geoboard square is equal to one half of the board, about how to measure the areas of these halves, and about whether two halves of a rectangle cut along diagonal and vertical lines have equivalent areas. Students use of phrases such as "First I thought . . ., but . . ." "We could use . . ." suggest their focused interaction with these ideas.

Ms. Richardson asks repeatedly how they might prove that what they have are halves. "Is this half? How do we know that? How could we prove that . . . ?" "If you can prove to us that this will make 8 squares . . ."

There are also examples of students building on each other's thinking, such as "You could do like PJ said . . . put that with this (i.e., put two triangles together), and make a square (i.e., a rectangle) . . . "

Ms. Richardson does a number of things both to support the belief that her students hold important mathematical ideas to explore and to encourage students to show respect for one another's ideas. She uses students' ideas to guide the whole group's initial exploration of how to prove equal area of halves: "Corey gave us an idea. We could actually measure it. . . ." She refers to students' "theories" and displays an interest in how they bear up to experimentation: "Do you think your theory might work? Are you beginning to wane on your theory?" Finally, she encourages students to give each other the time and space to think for themselves: "Let Marvin try to figure it out."

CLOSING
20 minutes

> ***Materials***
> Handout 4

Homework

5 minutes

Assigning Homework As you assign the homework (Handout 4 and Reading 2), preview the next two sessions during which participants will focus on the Intellectual Community Observation Guide. Participants will examine the relationship between the teacher's ongoing learning (about mathematics and children's mathematical thinking) and her capacity to support an intellectual community in the classroom in which new and unexpected mathematical ideas emerge. The homework reading introduces the idea that teachers might continue to learn mathematics while teaching.

Bridging to Practice

15 minutes

Reflective Writing This may be participants' first experience with such focused journal writing. Introduce it by telling them that the journal writing is designed to help them focus on the ideas in the course that seem particularly interesting for them and their school community. Note that the activity provides a structure for planning what they can pursue in order to move themselves and their schools along with these ideas.

Post the following questions, and invite participants to respond to them:

- *Pick an idea that came up today that you found particularly interesting. What is your current thinking about this idea?*
- *Where is your school now with regard to this idea?*
- *What are one or two things that you, as instructional leader, will pursue to move yourself and/or your school along with this idea?*

Bridging to Practice

At the end of this and every session, you will ask participants to take 15 minutes to write in their journals about the ideas from this session that were particularly salient to them. The purpose of the journal writing is for participants to have some time to reflect on the ideas discussed in the session and to articulate for themselves something that is important to them about these ideas. Writing also helps them make the bridge between these ideas and their own work as administrators.

School administrators have constant demands on their time and energy. Although it is essential for them to set these commitments aside in order to dig deeply into the ideas discussed in the session, it is also very important for them to think through how they might carry these ideas into their work as administrators. This is the intricate interplay between belief and action.

This may be the first time that some participants have engaged in reflective writing about their work with supervision, and you may wish to take a few minutes to introduce the topic, saying that the writing is not meant to be feedback on how the class went, but an opportunity for them to make note of ideas that were interesting to them and to push their own thinking forward. You may wish to collect copies of these reflections for your own use in understanding the thinking of the administrators in your class.

Your Own Journal Writing

Within one or two days of teaching this class, after you have had the chance to unwind from the class and perhaps talk with a colleague or friend about how it went, it will be important for you to set aside some time to write your own journal entry about the class. In general, we encourage you to write about the ideas held by the administrators in your group rather than writing about your own actions and how things worked out. In our experience, using the time to jot down what the participants in your class are thinking about will be much more useful to you as you begin to prepare for the next session. Some facilitators use the time when administrators are writing in their own journals to make preliminary notes for this journal writing.

What's In It for Us?

Take a few minutes to write about the following questions:

1. What do you need to know in order to be a good observer?

2. What do you, as an administrator, gain from conducting classroom observations?

Observation Guide

Students		
Focus Question	**Conjectures**	**Evidence from Classroom**
Math Content		
• What mathematical ideas are embedded in the lesson?		
• What makes this worthwhile mathematics?		
Learning		
• What kinds of mathematics sense-making are students doing?		
• What mathematical ideas seem to be confusing to students?		
• In what ways can you see that the students are developing their mathematical ideas over time?		
Intellectual Community		
• How are students showing respect for one another's ideas?		
• How do students use each other as resources as they make sense of mathematical ideas?		
• What evidence beyond raised hands do you have that students are engaged?		

Observation Guide

Teachers		
Evidence from Classroom	Conjectures	Focus Question
		Knowledge of Content
		• What does the teacher seem to understand about the mathematics?
		• What is the teacher's long-term mathematical agenda?
		• What does the teacher seem to understand about the development of children's ideas in this topic?
		Pedagogy
		• How does the teacher work with the sense the children are making?
		• How does the teacher work productively with students' confusion?
		• How does the teacher attend to all students?
		• How does the teacher adjust her teaching based on the ideas she hears from students?
		Facilitating Intellectual Community
		• How does the teacher support students in showing respect for one another's ideas?
		• How does the teacher set the tone so students see each other as resources for mathematical thinking?
		• What interventions does the teacher make to ensure that students' engagement has a focus on mathematical ideas?

Homework for Session 2

In preparation for Session 2, read *Learning Mathematics While Teaching,* by Russell, Schifter, Bastable, Yaffee, Lester, and Cohen (Reading 2 in the book of *Readings*). The authors look at three elementary teachers who "learn" mathematics in the context of their own teaching. As you read the article, consider the following questions:

1. What do the authors mean by "learning" mathematics while teaching?

2. What knowledge do teachers need to have to engage in the kind of learning of mathematics suggested by the authors?

3. Consider one of the three cases presented by the authors. How does being open to learning mathematics as this teacher was support the cultivation of a mathematical intellectual community in the classroom?

4. What seems appropriate and/or challenging about being such a learner, given what you know about the realities of teaching in your own school?

SESSION
2

Building Intellectual Communities in Mathematics Classrooms

In this session participants consider more deeply how they can develop a new eye for mathematics classrooms. Using the Intellectual Community Observation Guide, participants reflect on the nature of the intellectual community in the classroom they are observing and look at the qualities that exemplify such a community. Participants also look at how the interactions between teachers and students cultivate and sustain the intellectual community. We concentrate on the intellectual community dimension first in order to build on what many participants already attend to when they engage in classroom observations. These include such features as teacher questioning, student participation in class discussions, and opportunities for collaborative student work.

However, this session goes beyond simply observing procedural and interactive features to stress the interrelationship between these features and the mathematical content being explored. For many administrators, this focus will represent a shift in what they pay attention to in classrooms. Many administrators may recognize features that mark a supportive community, such as the level of trust between teacher and students or the level of comfort students feel taking risks. They may not, however, know how to attend to the ways that the classroom setting supports an environment of inquiry into substantive mathematical ideas. Yet, this environment of inquiry is essential in a standards-based classroom in which both the quality and depth of ideas explored are important. Administrators need to be able to discern not only whether the classroom has a supportive community characterized by trust and respect but whether the community fosters a fertile environment within which substantive mathematical inquiry can take place.

The activities in this session are designed to encourage participants to look beyond the surface features of the classroom. Participants describe the intellectual community in the classroom and consider how it supports the exploration of substantive mathematical ideas.

Overview for Session 2

OPENING page 48 20 minutes	**INTRODUCTION: WHAT IS AN INTELLECTUAL COMMUNITY?** Participants are presented with the main activities of the session and the connections to both previous and upcoming sessions. They are introduced to the Intellectual Community Observation Guide.
ACTIVITY 1 Video page 51 60 minutes	**LEARNING TO SEE A MATHEMATICAL INTELLECTUAL COMMUNITY** Participants view a videotaped clip, *Valentine Exchange.* Before watching the videotaped clip, participants work on the problem presented in the video. When watching the video, participants note how the teacher shapes the intellectual community in the classroom.
ACTIVITY 2 Homework Discussion page 58 25 minutes	**LEARNING MATHEMATICS WHILE TEACHING** Participants discuss the homework reading by Russell *et al.* (Reading 2). They reflect on ways teachers can be attentive to the mathematical ideas embedded in the curriculum and those raised by students to foster an intellectual community.
ACTIVITY 3 Video page 62 55 minutes	**WATCHING FOR MATHEMATICAL COMPLEXITIES** Participants view *Valentine Exchange* again. They develop an appreciation for the complexity of the mathematical terrain that teachers have to negotiate and the unpredictability of these negotiations. Participants also consider the importance of learning while teaching in order to support an intellectual community.
CLOSING page 67 20 minutes	**HOMEWORK** Participants read *What's All This Talk About "Discourse"?* by Deborah Ball (Reading 3). They also do an observation of an elementary math lesson in their schools or districts. They use the Intellectual Community Observation Guide to guide their viewing, and they write up a summary of their observations. **BRIDGING TO PRACTICE** Participants engage in journal writing to develop further some of their thinking from the session.

Big Ideas

Participants explore:

- mathematics classrooms as intellectual communities
- the benefits of learning mathematics while teaching for both teachers and students
- the unanticipated complexities mathematical inquiry can open up and the demands they place on teachers' mathematical and pedagogical skills

---

Materials

- ☐ nametags
- ☐ flip charts or overheads
- ☐ markers
- ☐ agenda
- ☐ Videotape 2, *Valentine Exchange* (DVD #1; Program 17; first viewing, 25:14–33:31; second viewing, 20:35–33:31)
- ☐ Handout 4, Homework for Session 2
- ☐ Handout 5, *Valentine Exchange* Mathematics Problem
- ☐ Handout 6, Discussion Questions for *Valentine Exchange*
- ☐ Handout 7, Overall Observation Guide
- ☐ Handout 8, Intellectual Community Observation Guide
- ☐ Reading 2, *Learning Mathematics While Teaching,* by Russell *et al.*
- ☐ Handout 9, Homework for Session 3
- ☐ Handout 10, Pre-observation Conference Questions
- ☐ Reading 3, *What's All this Talk about "Discourse"?* by Deborah Ball

---

Preparation

- Prepare an agenda for the session on an overhead or on flip chart paper with approximate times for each activity noted.
- Write the prompts for the Bridging to Practice on an overhead or on flip chart paper.
- Review who your participants are and why they are interested in taking the course. If you are collecting their writings, read through them very carefully to gain a sense of their understanding of what the course is about and their views of classroom observation and supervision in general. Paying attention to this early in the course will help you tailor the course to the needs of your particular group by selecting the discussion questions your group would most benefit from considering.
- Review the Intellectual Community Observation Guide. You want to have a clear sense of the different aspects of the intellectual community of a classroom so that you can help participants develop an image of an intellectual community.
- Work through the *Valentine Exchange* problem of Activity 1 yourself. As you do so, imagine the responses of students as they puzzle over how to systematically figure out the number of valentines there would be. Perhaps ask a child to do the problem. It will help you understand more fully the complex mathematical terrain that Ms. Olivez's modeling of the problem opened up.

PREPARATION

- Read the article *Learning Mathematics While Teaching* by Russell, Schifter, Bastable, Yaffee, Lester, and Cohen (Reading 2) for Activity 2. Pay special attention to the three cases presented. It is important for you to have a good sense of what the authors mean when they refer to learning mathematics while teaching. Because the notion of learning while doing may seem unusual or counterintuitive, it might help you to think about situations in which one can learn while being actively engaged in one's daily work. What, for example, might you learn about teaching and mathematics when you observe a mathematics lesson or engage in a post-observation conference with a teacher?
- Preview at least two times the videotaped clip *Valentine Exchange* for Activities 1 and 3. View the first version, 25:14–33:31, which has a short section from the beginning of the class edited out. Use the Intellectual Community Observation Guide to see how it draws attention to the class as an intellectual community. Pay particular attention to how Ms. Olivez creates an atmosphere of inquiry in the classroom and how the students work with one another to solve the problem. Remember that the first time participants watch the video, they will not see the first few minutes of this class. In the full version, 20:25–33:31, pay particular attention to how Ms. Olivez models the *Valentine Exchange* problem and how that led students to investigate the problem in certain ways.
- As we advocate throughout this course, plan to leave enough time in this session for participants to view the videotaped clip two times. You may want to refer to the section on *Viewing Videotaped Clips* in the **Introduction** (p. 9) to remind yourself of the rationale that underlies this recommendation.

OPENING
20 minutes

Materials
Name tags
Agendas
Flip chart paper
Markers
Handouts 7 & 8

What Is an Intellectual Community?

Before starting the session, post the agenda in a place where everyone can see it.

5 minutes

Introducing the Session Because this is only the second session of the course and you may have a few newcomers, invite everyone to reintroduce themselves. Ask participants to include their roles and school districts in their introductions.

Explain to participants that a main focus of this session is on learning to see mathematics classrooms as intellectual communities. Remind them of the overall purpose of the course and the two strands that will be explored:

Developing a new eye for mathematics classrooms

Rethinking administrators' talk with teachers

Point out to participants that one aspect of developing an eye for mathematics classrooms is attending to the intellectual community.

Explain that in addition to considering what an intellectual community is, participants will be exploring both the role of teacher as learner within such communities and the complex mathematical terrain that such communities can open up.

10 minutes

Individual Writing and Small Group Sharing Ask participants to write what they think of when they hear the phrase *intellectual community.* You might say something like the following:

> *The phrase,* intellectual community *is often used to describe standards-based mathematics classrooms. Take a few minutes to jot down some thoughts on what an intellectual community in a mathematics classroom is, noting especially what you think both the teacher and the students do.*

After a minute or two, have participants gather in groups of three or four and share their thoughts. Circulate among groups to hear the ideas being shared. Request that a member of each group write the key points on a sheet of flip chart paper. Keep these papers in view throughout the session so that both you and participants can refer back to them as needed.

This is not a time for an extensive discussion; it is a time for participants to share their thoughts and for you to get a sense of participants' ideas of an intellectual community. Knowing how participants interpret the phrase *intellectual community* will help you shape the discussions in the activities that follow.

5 minutes

Introducing the Intellectual Community Observation Guide Distribute the Overall Observation Guide and the Intellectual Community Observation Guide. Remind participants of the general structure of the Overall Observation Guide, which is divided into three sections: Math Content, Learning/Pedagogy, and Intellectual Community.

Overview

These notes are intended to prepare you to identify more readily the core ideas of each activity and to help you direct the discussion to them if they have not come up. As you prepare for each activity, keep in mind that the discussion questions presented are meant to stimulate such discussions. A critical component of facilitating discussion is to listen closely as participants consider these questions so that you can understand where they are with regard to them. One of your tasks as facilitator is to ask pertinent follow-up questions that offer participants the opportunity to dig more deeply into the important ideas of the activity.

Group Sharing of Individual Writing

In this session, participants focus their attention on the *intellectual community* of the mathematics classroom and the ways it contributes to student learning. The brief warm-up exercise in which participants write what they think of when they hear the phrase *intellectual community* starts them thinking about intellectual communities.

As participants share their thoughts on intellectual communities, you and they will hear one another's ideas on the topic. Some people may focus primarily on the process features of a classroom, emphasizing students' opportunities to share strategies or express confusion. Others may focus on the teacher's style and tone of voice, emphasizing the culture or atmosphere of the classroom. From these comments, you can get a sense of what impression the use of the Observation Guide in the last session has had on their thinking.

The notion of "intellectual community" as used in this course emphasizes the *intersection* between the behaviors, practices, and values of a standards-based classroom and the content being explored. The emphasis is on how classroom behaviors, practices, and values support the learning and reasoning needed to make sense of complex mathematical ideas. The "community" component highlights the importance of an atmosphere of respect, trust, and commitment to supporting everyone's learning while the "intellectual" piece stresses the continual responsibility to the content to be learned.

The Intellectual Community Observation Guide

Participants look at the Intellectual Community Observation Guide, which presents guiding questions for participants when attending to the learning climate in the classroom.

The first set of questions rests on the key principle that mathematical ideas are constructed by individuals (rather than a set of rules and procedures created by an outside authority to be absorbed by learners). Thus, in a standards-based classroom, it is essential that students be supported in their efforts to develop understandings of mathematical ideas. Behaviors such as attentively listening to one another's thinking, posing questions about mathematical ideas, and sharing provisional thinking can help in this development. (Provisional thinking refers to emergent understandings of complex mathematical ideas such as why the product of two fractions is a smaller fraction. Students need to feel comfortable expressing their tentative understandings of concepts.)

The second set of questions is grounded in a central tenet of the theory of the social construction of knowledge: that the process of articulating one's own thinking and of building on the thinking of others is a key component of making meaning. Thus, this set of questions draws the observer's attention to how students serve as resources for one another and build on one another's ideas.

The third set of questions directs the observer's attention to the nature of students' engagement with mathematical investigation. As part of their focus on the nature of the intellectual community of the classroom, observers might also want to consider the characteristics of the students in the class (gender, race, language, and so on).

ACTIVITY 1
Mathematics Exploration
and Video
60 minutes

Materials
Videotape 2
Handout 5

Learning to See a Mathematical Intellectual Community

10 minutes

Exploring the Mathematics Tell participants that they will view a section of a 15-minute videotaped clip on patterns and numerical relationships. Explain that before they watch the video, they will take a few minutes to do the actual problem in the videotaped clip. Present to participants the two main reasons for doing the problem: participants will be able to focus on the features of an intellectual community better in the first viewing and concentrate on the mathematical issues that arise in the second viewing.

Pass out *Valentine Exchange* Mathematics Problem (Handout 5).

> *There are 24 students in Ms. Olivez's class. If each student gives a valentine to everyone else in the class, how many valentines will there be?*

Encourage participants to solve the problem using more than one strategy. Suggest that they use a strategy involving numbers and one that makes use of pictures or graphics. Give participants about 5–7 minutes to work on the problem by themselves. If it appears that some people are not able to solve the problem by themselves, you can suggest that they confer with a neighbor.

10 minutes

Sharing Approaches to Problem Solving After participants have had a few minutes to work on the problem, ask for someone to share his or her strategy.

It is likely that the first strategy offered will be to multiply 24 by 23. If this is the case, you might ask the participant to explain how he or she knew that multiplication could solve the problem and that 24 and 23 were the numbers to multiply.

After the first person has shared his or her strategy and the group has had an opportunity to discuss it, ask for a different strategy. After participants have shared a couple of different strategies, you might ask participants to share their process for solving the problem: how they defined the problem and developed a strategy for solving it. For instance, did they start by making a visual representation of the 24 students? Did they realize right away that the problem could be solved by multiplying?

The intent of asking participants these questions is to help them recognize the importance of organizing and problem-solving skills to making sense of a problem and figuring out what one needs to do in order to solve it.

5 minutes

Introducing the Videotaped Clip Tell the group that they will now watch the video with an eye on the intellectual community in the classroom. Explain that a short section of the video has been edited out for this first viewing.

Remind participants that the central purpose for viewing and discussing classroom episodes is not to critique the teaching depicted there. Although they might be tempted to do this, stress that they simply will not know enough to make valid judgments about the teaching they observe. Teaching is complex, and from the perspective of an outside observer, there are always other moves that might have been made. An outside observer who sees only a short excerpt of a lesson, such as what participants will see in these videos, will not know enough about the backgrounds of the students in the class, the context of the lesson, or the teacher's thinking to make valid judgments about whether the teacher's decisions were good ones. Tell them that for this viewing they will focus on the questions in the Intellectual Community Observation Guide.

Give them the following background information about the classroom:

- This is a fourth grade Spanish bilingual classroom.
- Students explore patterns and relationships through the *Valentine Exchange* problem.
- The 15-minute videotape has been edited to show portions of the introduction, small group work, and whole group discussion of the investigation. Occasionally Ms. Olivez, the teacher, is heard in a voice-over reflection. In this viewing, about two minutes from the beginning of the class have been edited out. Because of this editing, classroom dialogue will start quickly. Prepare participants for this.
- Prior to the beginning of the clip, the teacher introduced the problem.

15 minutes

Watching the Videotaped Clip Show the first version of *Valentine Exchange* (25:14–33:31), and then give participants a few minutes to write down notes on aspects of the intellectual community that they noticed in this episode.

20 minutes

Discussion of the Video with a Focus on Intellectual Community. In order to ground the discussion in the students' mathematical thinking, start the discussion with the first question:

What mathematical ideas do you see students exploring?

Because participants saw the video only once, they might not be as able to recall what various students actually said. You may need to prompt them by mentioning a particular student.

After participants have had some time to consider the mathematical ideas students were exploring in this episode, you can shift the discussion to consider how the intellectual community supports students serving as resources for one another. You can ask:

In what ways did students use each other as resources as they made sense of mathematical ideas?

Because participants will not yet have seen the beginning of the class, during which Ms. Olivez models the problem and some ideas for solving the problem are first generated, they will be limited here to considering a few interactions among students. If the discussion is slow, you can ask participants about particular events in the classroom. For example, ask participants to comment on this interaction between two boys: the first boy paired off students to come up with an answer of 48 and the second responded that the answer was probably much larger.

You will also want to draw attention to the ways Ms. Olivez helped students articulate their thinking. You might ask the following question from the Intellectual Community Observation Guide:

What evidence is there that Ms. Olivez invited students' tentative thinking?

Explain to participants that in the next activity, they will explore how teachers who teach in standards-based classrooms can continue to be "learners" of mathematics while teaching it. Note that this discussion will serve as a backdrop for the second viewing and discussion of the classroom episode that follows.

Doing the Mathematics

In this activity, participants solve a short mathematics problem taken from the videotaped classroom episode. Because this is the first time participants solve a mathematics problem, you should review the reasons for this activity in the "Doing Mathematics" subsection of the *Pedagogical Considerations* section in the **Introduction.** On page 8, two key reasons for doing the short mathematics problem are presented for you to share with the group.

The problem has been deliberately worded to encourage participants to see it as a multiplication problem: 24 students times the 23 valentines that each would need to bring. (They do not have to bring a valentine for themselves.) This was done primarily for efficiency; that is, we want participants to be able to solve the problem in the ten minutes provided. Still, some participants may not immediately see that the problem can be solved by multiplying the number of students in the class and the number of valentines that each student would need to bring. They may end up getting stuck (as did some students in the videotaped clip) on how to keep track of all of the exchanges of valentines. Others may simply not realize that each student needs to bring only *23,* and not 24, valentines.

Synopsis of the Videotaped Clip

This synopsis is of the full version of the videotaped lesson. The beginning of the lesson is edited out of the first viewing. We highly recommend that you watch both versions of the tape carefully to be clear about what portions have been edited out in the first version. We also recommend that you review the guidelines for setting a tone for viewing videotapes (see notes in the **Introduction** of this course). The first (edited) version of the lesson helps participants focus more readily on the intellectual community without getting distracted by Ms. Olivez's modeling of the problem. (This portion of the lesson will be discussed after the second viewing.)

(This part is edited out of the first version.) The clip starts with Ms. Olivez recounting the valentine party that they had the previous day as an introduction to the problem. A voice-over explains that Ms. Olivez is going to use valentines in a lesson on patterns and functions.

Ms. Olivez has one student, Gina, come to the front of the room. She hands her a valentine and takes a valentine for herself. She then exchanges valentines with Gina, asking the class, "If I gave Gina a valentine and Gina gave me a valentine, how many valentines have we given, have we exchanged?" Chelsea answers two and Ms. Olivez asks her to explain. Ms. Olivez then has three students come up and exchange valentines. She asks how many valentines have been exchanged and Josephina answers six since each person exchanges with each of the other two. Winston asks (in Spanish) why it has to be six. Ms. Olivez repeats the question in English. Gina explains that a student "can exchange as many times as you [the student] want." Denri agrees and points out that he could exchange with Ms. Olivez twice. In response, Ms. Olivez imposes a constraint on the activity: "But what if we were going to exchange one valentine at a time." Gina then gives an answer of four. Students at the front model the exchanges again, showing that the number of exchanges is six.

Ms. Olivez asks whether this number would change if there were four people exchanging instead of three. One student says that the number of exchanges would be eight. He goes through the exchanges one by one to discover that eight exchanges are not enough for everyone to have exchanged with everyone else. He says, "It could be more." Denri, appearing to have visualized the exchanges in his head, says that it could be twelve. The students in the front of the class model the exchanges to show that the number is indeed twelve. Ms. Olivez asks whether anyone has a different answer and strategy. Armando comes to the board and draws four stick figures, labeling each with the first initial of one of the students at the front of the room. He then makes a table beneath the figures to indicate all of the exchanges that would take place among the four people.

(This part is included in the first version.) Ms. Olivez poses the problem that she would like the students to explore. She says in Spanish, "We have 24 students in this class and everybody has a card." The voice-over says, "Ms. Olivez then poses a challenge. If each student gives a valentine to everyone in the class, how many valentines will there be?" Ms. Olivez invites students to work in the grouping that they feel most comfortable: with their partners, in groups, or by themselves.

Two students develop a strategy for solving the problem: "I think the big ones should be people. . . ." Ms. Olivez checks in with three different groups or individual students. In the first instance, Winston and two other students have started to solve the problem by having pairs exchange with each other. When Ms. Olivez joins the group, they explain to her that there are 24 students and 24 valentines. They have built a representation of the 24 students using pattern blocks paired in two rows of 12 each. One student explains that as the pairs exchange with each other, there will be 24 exchanges and then says that number can be doubled to get 48 exchanges. Another student recognizes that there will be many more exchanges because everyone has to exchange with everyone else and not just with one other person: "Greg has to exchange with Louisa and then Louisa has to exchange with this person and this person has to exchange with this person and Louisa has to give Greg one so 48 is probably not the answer. It's probably much larger." Ms. Olivez then asks them whether they would consider changing the 48 and the student responds yes.

Denri tries to simplify the problem by solving it for 12 instead of 24 students. He has made a model using pattern blocks and cubes. One cube represents one student, and a hexagon represents each time a student exchanges a card. He notes that when each student has made twelve exchanges, everyone has exchanged with everyone else.

Two students are also puzzling about where to go beyond doubling 24 to get 48. They have somehow figured that they would have 9 additional exchanges.

Ms. Olivez checks in with Gina who is working with Cuisenaire® rods and has made a 12-by-12 grid on piece of paper. She also was working on solving for 12 students. She explains that the rows represent the students and the columns the cards. This is her way of recording who has exchanged cards with whom: "So when this one already exchanged with number 12, I can put a check on the number one." In explaining her strategy to Ms. Olivez, it appears that she might be renaming the column as representing students instead of cards.

A whole group discussion follows the independent/group work. Ms. Olivez asks students to share their thoughts on the activity and the process. Denri thinks that he and Adam would have come up with the answer because they were cooperating. Gina got confused because she did not know how to start out. Another student shares his strategy: he drew a picture of the seating arrangement in the classroom, divided the exchanges in some way among the students at each table, and then added those exchanges up, getting 105. Chelsea tried to track all of the exchanges as the cards moved around the room: "I said, well, okay, I'm going to get one from Denri and give it to Adam and from Adam and give it to Denri around the class but then I got mixed up and had to start all over. Really hard." Josephina shares her strategy but expresses surprise at how large the answer is: "If you had 24 people in your class you will give a card to each person except yourself.

So, you will give 23 cards out. So, 23 cards . . . so, 23 times 24." As she reads her answer, she changes her statement to a question: "And I came up with 552?"

Discussion of the Videotaped Clip

Participants consider three questions in the first viewing of the videotaped lesson. Be sure to keep the focus in this portion of the discussion on the mathematical *ideas.* Some participants might offer more general comments such as, "They are learning how to find patterns." You might then follow a statement like this with the question "What is mathematically significant about patterns?" Or, participants may note that students are exploring addition or multiplication. Again, probe a bit by asking what specific concepts relating to addition or multiplication students are exploring.

♦ **What mathematical ideas are students exploring?** This clip shows students working to think systematically about a multiplication problem. They use various models to represent and keep track of their thinking. This can indicate the beginning of a process for finding a general rule that will work for any step number of a series without finding all the terms in between. Two challenges that students face include representing the problem so that they can account for all of the exchanges and simplifying the problem so that they can solve it in smaller chunks.

Winston and two other students begin by finding out how many exchanges there will be when pairs give each other valentines. They appear to be working on how to represent the exchanges so that they can accurately count them.

Both Denri and Gina use a strategy for finding how many valentine exchanges there would be for half the number of students. The model that Denri uses does not account for the fact that a student does not exchange with himself or herself. The grid that Gina develops has the potential for helping her realize that one does not exchange with oneself, but it seems that she has not figured out how she can use the grid she has developed. We do not see how she actually filled in the grid. The complexity of finding the number of exchanges for half of the students and doubling it as well as the reason it does not work to solve the problem is discussed in the **Facilitator Notes** for Activity 3.

At the end of the videotaped clip, Josephina gives a clear articulation of the logic she has applied to the problem. She explains that there are 24 students, that each of the 24 students will give her a valentine, and that she will give each of the other 23 students in the class a valentine. Thus, the solution is 24×23, which she computes to be 552. Note that Josephina counts the number of valentines each person would bring to give one to everyone else rather than counting the number of exchanges that would occur if everyone had one card and exchanged that card with everyone else.

♦ **In what ways did students use each other as resources as they made sense of mathematical ideas?** Because participants will not have yet seen the beginning of the video where some ideas for approaching the problem are first generated, they will be limited in the interactions to consider. The interaction between the two boys previously described is an interesting one to discuss with participants. You might ask participants to speculate on students' thinking as they looked for solutions. As participants present their conjectures about the exchange, ask them to cite specific evidence.

♦ **What evidence is there that Ms. Olivez invites students' tentative thinking?** Participants might mention a number of examples where Ms. Olivez helps students articulate what they are doing and has them describe to her the models they are using. You may need to encourage participants to be specific about how Ms. Olivez's action helps students articulate their thinking. For instance, someone may say that she asks them to tell her what they are doing. You may have to push this a little further to help them notice what Ms. Olivez actually asks: "Oh, these are the children, the students?" or "And when you have counted all the squares, what does that mean to you?" You may need to help participants cite specific evidence to support the general observations or conjectures. This is an important habit to develop as participants move toward considering collaborative inquiry as a mode of supervision in later sessions of this course.

Other Considerations

Some participants may comment that Ms. Olivez does not seem to pursue actively the ideas of students who appear to be on the wrong track or come up with the incorrect answer. If this concern comes up in your group, remind participants that the episode has been edited, making it difficult to draw valid conclusions about the classroom teacher from what they have seen. Further, because we do not have access to Ms. Olivez's thoughts about the pedagogical moves she chose to make, we should refrain from passing judgment. However, if such issues come up, you may wish to illustrate for the class how to generate questions for Ms. Olivez that would give insight to her thinking about the mathematics and her students' thinking during this class session. Examples of such questions might include "What do you think Gina understood about the problem, and where might the soft spots in her thinking be?" or "What do you think Josephina's surprise at the number of valentines indicates?" Such questions are examples of collaborative inquiry into student thinking that will be explicitly addressed in Session 5.

ACTIVITY 2
Homework Discussion
25 minutes

Materials
Handout 4
Flip chart or overhead
Markers

Learning Mathematics While Teaching

10 minutes
Whole group

Discussing the Russell *et al.* article Tell participants that they will now consider one way a teacher can foster an intellectual community in the classroom: by being a learner himself or herself.

Have a flip chart or overhead ready, and ask participants the first question from their homework assignment:

What do the authors mean by "learning" mathematics while teaching?

As different people share what they understood this to mean, note their ideas on the flip chart or overhead. Pay attention to how people have made sense of the ideas the authors present. Make sure that participants have picked up on the authors' contention that learning mathematics while teaching is not merely a remedial strategy for teachers who are new to ideas of reformed practice, but an ongoing and integral part of the new pedagogy. If it seems as if participants are stuck on the idea as a remedial one, you will want to ask a question like the following:

Many of you focused on what mathematics teachers need to learn in order to teach well. How would teachers who are already comfortable with the mathematics continue to be learners in their classrooms?

Your goal here is to make sure that participants have grasped the authors' meaning of being a "learner" while teaching mathematics and especially that they have not misunderstood it to mean that most teachers are lacking in content knowledge and therefore have to learn "on the job."

When participants seem to have a good grasp of this core idea, ask the second question from the homework:

What knowledge do teachers need to have to engage in the kind of learning of mathematics suggested by the authors?

You will want to highlight the authors' assertion that teachers need to be open to learning and be curious about mathematics and how children learn it.

5 minutes
Small group

Discussing One of the Cases Ask participants to turn to a neighbor and discuss the following question for about 5 minutes:

> *Consider one of the three cases presented by the authors. How does being open to learning mathematics as this teacher was support the cultivation of a mathematical intellectual community?*

10 minutes
Whole group

Learning and Teaching Mathematics After participants have had a few minutes to discuss the question with a partner, ask the pairs to share key points of their conversation. Listen for how participants are developing their understanding of the classroom as an intellectual community.

As a way of drawing out participants' own experience, ask the final question from the homework:

> *What seems appropriate and/or challenging about being such a learner, given what you know about the realities of teaching in your own school?*

Tell participants that in the next activity, they will use this idea of teachers being "learners" to understand how mathematical investigations in classrooms that are intellectual communities can open up mathematical complexities for both students and teacher.

FACILITATOR'S NOTES
ACTIVITY 2

Overview
Creating and supporting an intellectual community requires that teachers recognize students' thinking and make pedagogical moves that are appropriate to the situation. These moves might consist of listening carefully to students' ideas, asking questions that can help students articulate provisional thinking, or helping students stay focused on mathematical ideas.

To support an intellectual community, teachers have to be "learners" themselves; that is, they have to "hear" the mathematical ideas that students raise and be open to exploring these ideas themselves. This way of thinking ties closely to the concept of *generativity* that was presented in the **Introduction.** When teachers focus on how their students are making sense of the mathematical concepts being taught, they set the stage for increasing their own understanding of these concepts and of the ways students think about them. This idea of generativity will be addressed more fully in Session 5.

In the article *Learning Mathematics While Teaching,* Russell *et al.* explore three different ways teachers can learn mathematics while teaching. They maintain that when teachers are able to be learners themselves—when they are working at learning more about the mathematics they teach—they are much better prepared to facilitate an intellectual community and learning experiences that can ensure student learning.

Discussion Topics for the Russell *et al.* article

♦ **What do the authors mean by "learning mathematics while teaching"?** Russell *et al.* identify the mathematics classroom as an important context in which teachers can extend and deepen their understanding of the "varied, complex, and context-dependent" understandings of mathematics; the ways children develop mathematical understandings; and the teaching situations teachers must negotiate. More than the acquisition of new knowledge, this classroom-based learning involves "a more ongoing and gradual process in which understanding of familiar content is deepened as one makes new connections and distinctions" (p. 24).

♦ **How would teachers who are already comfortable with the mathematics they teach continue to be learners in their classrooms?** Teachers who are focusing on how students are making sense of the mathematical ideas under discussion are gaining insights into the range of ways that students tend to think about these ideas. In the process, these teachers' own understanding of the mathematical concepts can be stretched and deepened. These teachers are also learning what pedagogical moves are likely to stretch students' mathematical thinking.

♦ **What knowledge do teachers need to have to engage in the kind of learning of mathematics suggested by the authors?** Russell *et al.* note that teachers need to view themselves as ongoing adult learners and their classrooms as settings in which they can learn. They need to be, or to have been, engaged in substantive experiences of mathematics learning as adults. They also need to share the following three assumptions: "1) learning about mathematics occurs when one is immersed in problem solving; 2) an important aspect of mathematical thinking is identifying, describing, and testing patterns and relationships; and 3) understanding the mathematics better has implications for their pedagogical decisions" (p. 29).

♦ **Consider one of the three cases presented by the authors. How does being open to learning mathematics as this teacher was support the cultivation of a mathematical intellectual community in the classroom?** Each of the cases offers a different aspect of learning mathematics while teaching it. In the first case, a teacher discovers the complex mathematics behind what initially seemed to be a straightforward problem.

She learns through this experience the importance of exploring the mathematics for herself and of being able to design a problem that fits more with what is appropriate for second graders. This particular teacher gains a deeper appreciation of the more complex mathematics that lies behind the elementary curriculum. This is illustrated in the classroom episode for this session: Ms. Olivez is open to exploring the mathematics for herself and so becomes more skilled at facilitating situations in which students get into more complex mathematical terrain themselves.

The second case portrays a teacher who sees an alternative way of conceptualizing subtraction by paying careful attention to students' representations and strategies. Being open to not only learning about but also learning *from* students' representations helps teachers avoid a perspective of "one right way" and makes teachers more receptive to multiple solutions and strategies. At the same time, teachers need to examine the validity and mathematical integrity of various representations, which often require a careful exploration of these representations. Otherwise they can fall into the opposite perspective of accepting any strategy as equally valid.

In the third case, a teacher discovers that some of her students' understandings of the base-ten system are more fragile than she assumed. By looking more carefully at what is problematic for students, she develops a deeper understanding of the complexity of coordinating multiple units and appreciates more fully how challenging it can be for students to grasp this concept. In order to facilitate and support an intellectual community in the mathematics classroom, a teacher has to be able to look beyond students' reasoning and explanations and pinpoint concepts with which they might be grappling. In the video, several students were trying to simplify the problem by solving it for twelve students and intending to double it. In order to understand what these students are attempting to do, Ms. Olivez needs to continue to explore for herself the structure of the operation of multiplication.

♦ **What seems appropriate and/or challenging about being such a learner, given what you know about the realities of teaching in your own school?** This question is intended to move the discussion from just examining the ideas in the article to considering the experiences that participants bring with them. As participants consider the idea of teachers being "learners," it is important for them to think about what feels right about this stance but also what might feel challenging about making it a reality in their schools. In this part of the discussion, encourage participants to consider the implications of this stance for their own schools.

ACTIVITY 3
Video
55 minutes

Materials
Video 2
Handout 6
Flip chart or overhead
Markers

Watching for Mathematical Complexities

15 minutes

Second Viewing of the Videotaped Clip Explain to participants that they will now watch the unedited version of the videotaped clip (20:15–33:31). Note that the beginning segment of the video, which they did not see before, shows how Ms. Olivez chose to set up the valentine exchange problem for the class.

Distribute Discussion Questions for *Valentine Exchange*—Second Viewing (Handout 6). Tell participants that they should pay particular attention to how even a good teacher might get into complex mathematical terrain that she had not anticipated.

40 minutes

Discussing the Videotaped Clip Start the discussion with the first question:

What is happening at the beginning of the class? What do you think is the initial problem that Ms. Olivez posed? What are some different ways of interpreting what she said?

Because participants have already done the mathematics problem, it may not be as evident to them that it is potentially ambiguous. You may need to say something more specific such as:

What do you think Ms. Olivez is asking the students to find?

What do you think is the difference between finding the number of valentines and the number of exchanges?

What are the different ways of interpreting the word exchange?

You might want to give about 15 minutes or so to this portion of the discussion, particularly if it seems appropriate to model the difference between distributing valentines and tracking exchanges.

You will then want to shift the focus to have participants consider how the setup of the problem shaped the ways in which students approached the problem. Ask the next question on the handout:

How does the modeling and discussion at the beginning of the class seem to affect the strategies and approaches students select to try to solve the problem?

Here, you will want to draw out the power that modeling has on shaping the ways students think about and approach the problem. For example, the setup emphasized a one-to-one exchange rather than a serial exchange (student one exchanges with everyone, then student two exchanges with everyone, and so on). Many students worked with this one-to-one exchange model rather than a serial exchange.

Make sure that you leave at least 10 to 15 minutes to discuss the final questions:

> *What are some of the mathematical complexities that the teacher and the students encounter as the activity progresses?*
>
> *How does the intellectual community help bring to the surface these mathematical complexities? How might this become a constructive development?*

Especially because participants will have considered how the setup shaped the subsequent exploration, it is important that you emphasize here that the complexities that the students and Ms. Olivez encountered are inherent in the mathematics and not solely a function of her modeling. You will want to be mindful of participants who might offer a quick fix: "If only she had . . ." Be sure to draw attention back to the *mathematical* complexities that this type of problem can open up.

Overview

It will be useful for you to read through the video synopsis at the beginning of the **Facilitator Notes** for Activity 1 again, paying attention to the short section at the beginning of the lesson that was edited out for the first viewing of the videotaped clip. It was edited out initially to help participants stay focused on the nature of the intellectual community in the classroom rather than focusing on the potentially complicated way in which Ms. Olivez modeled the problem with the class.

Discussion Questions for the Second Viewing

♦ **What is happening at the beginning of the class? What do you think is the initial problem that Ms. Olivez posed? What are some different ways of interpreting what she said?** At the beginning of the class, Ms. Olivez models the problem that she would like students to explore on their own. There is some confusion as to whether she is asking students to *count* valentines or cards, as participants were directed to do in their initial mathematics exploration, or to *exchange* them. In her modeling, Ms. Olivez uses a fixed number of valentines and has students count the number of "exchanges," or how many times each student exchanges with all of the other students in the group. Both the card model and the exchange model result in the same answer, which can be expressed algebraically as $n(n - 1)$, where n is the number of students. The card model, however, seems to lead more easily to a multiplication solution.

Another possible source of confusion is in understanding how to count the exchanges. Although it does not appear that students encountered this confusion, it is possible to have different interpretations of what counts as an exchange. In modeling the problem, Ms. Olivez referred to an exchange each time a card was passed to another person. However, one can also interpret an exchange as being a reciprocal transaction—both people giving each other a card, similar to a handshake. When two people shake hands, there is only one handshake. Thus, if the exchange were understood as being a reciprocal event, like a handshake, there would be half as many *exchanges* as *cards exchanged*. This can be expressed algebraically as $\frac{n(n-1)}{2}$. Because Ms. Olivez modeled the problem as counting exchanges, some students might be confused as to what constitutes an "exchange."

♦ **How does the modeling and discussion at the beginning of the class seem to affect the strategies and approaches students select to try to solve the problem?** The way Ms. Olivez models the problem has likely influenced the way students approach the problem. We provide two examples here. First, she structures the problem so that students focus on the one-to-one exchange process. This process may be closer to what students actually do when they exchange valentines: Each gives a valentine to the person and the person gives one back to him or her. When students try to devise models and accounting systems for keeping track of the exchanges, many of them continue to focus on the one-to-one exchange process. They do not, for example, construct a model that considers all of the exchanges of one student before considering those of the next student. This may have been deliberate on Ms. Olivez's part because she may have wanted to create a situation similar to an actual situation—the seeming chaos of a valentine exchange among 24 students—and she wanted to give students the opportunity to construct an accounting system of their own.

Secondly, because she modeled the problem with each student having only one card, she drew attention away from thinking about the problem in terms of *cards*. She may have wanted students to focus on the number of *exchanges* rather than the number of cards. However, keeping track of exchanges is more difficult than keeping track of cards because the exchanges are less tangible than cards.

♦ What are some of the mathematical complexities that the teacher and the students encounter as the activity progresses? This classroom episode depicts the students encountering a number of mathematical complexities that are inherent in the problem. Some students are struggling to devise accounting strategies to keep track of exchanges. For many students, figuring out what is being asked in a problem and then systematically organizing the problem in a way to solve it is challenging. Ms. Olivez quite possibly did not anticipate how challenging it would be for students to keep track of exchanges, particularly in terms of seeing the growth pattern of the exchanges as they are linked to multiplicative properties.

We see evidence of this struggle when we look at the students who attempted to make the problem more manageable by working with half the number of students and then doubling their answers. Halving or doubling may be a strategy that can work in some mathematical situations, but when dealing with a multiplicative situation like this one, halving and doubling would *not* work because when the number of students increases, so would the number of valentines each student brings. There are many more valentines or exchanges added when you double the number of students. Ms. Olivez probably did not anticipate that students would decide that working with the smaller number would be a useful strategy, nor did she anticipate how complex it would be to figure out why that strategy is *not* valid in this situation. We can consider this strategy in more detail in case it seems appropriate to pursue this in the session.

When Denri decides to solve the problem for twelve students, intending to double his answer, he demonstrates his understanding that there would be more valentines for more students and that multiplication is involved somehow. However, doubling his answer of 132 valentines does not yield the correct answer. In Denri's modeling, there are eleven valentines given away by each of twelve students, which can be expressed as $11 \times 12 = 132$. In doubling the number of students ($11 \times 24 = 264$), the answer of 264 reflects only the increased number of students, not the increased number of valentines each student would need to bring. However, each student has to bring 23 valentines to have enough valentines for everyone in the class: $23 \times 24 = 552$.

If we think of multiplication as the number of items in a group multiplied by the total number of groups, when the number of students increases, not only does the number of groups increase (i.e., number of students), but so does the number of items in each group (i.e., number of valentines students need to bring in). Doubling the number of students from 12 to 24 thus increases not only the number of "groups" but also the number of items in each group.

Denri found that, for twelve students, there would be a total of 132 valentines and realized that doubling the students to 24 would result in too few valentines. How might he work from this point to find the right number of valentines? One strategy he might develop is to add more valentines to each student's supply:

$11 \times 24 = 264$

$12 \times 24 = 288$

$(11 \times 24) + (12 \times 24) = (11 + 12) \times 24$

$23 \times 24 = 552$

Although students rarely use such efficient notation, this is the sort of sense-making elementary students tend to engage in when they are in classrooms in which an intellectual community of collaborative math learners exists. This type of problem solving is important for students to engage in as they develop their algebraic reasoning skills throughout elementary school. Even though elementary students are not likely to fully understand and incorporate these skills into their mathematical processing immediately, such reasoning skills constitute an essential foundation to students' algebraic thinking and need to be developed gradually, over the elementary school years. When students begin their study of algebra in middle school, those who have developed these sense-making and problem-solving skills will be better prepared to undertake formal algebra. Participants in the course need to recognize how conceptual understanding builds slowly through repeated investigations and exploration of similar ideas.

♦ **How does the intellectual community help bring to the surface these mathematical complexities? How might this become a constructive development?** This question is intended to help bring the focus back to how the intellectual community in a classroom helps shape the mathematical content that is explored. In the *Valentine Exchange* episode, the intellectual community allowed a number of complex issues about the mathematics underlying the problem to surface. Students experiment with a range of materials to develop models and strategies to solve the problem. They uncover complex features of multiplication and addition. The strategy of finding the answer for twelve students and then doubling it, for instance, opens up issues of how multiplication and addition work (or do not work). The students who come up with an answer of 48 when they have pairs exchange valentines have to rethink what did not work in their modeling of the problem.

A classroom like this one, in which students are exploring concepts and working to make sense of the mathematics, provides teachers with opportunities to pick up on any number of issues and do a more extended exploration. In this classroom, for instance, Ms. Olivez might want to focus on the differences between addition and multiplication or the effects of using a particular model or strategy to solve the problem. Students could share how they rethought the problem when their initial strategies did not work and what they discovered in the process of this rethinking. It is important to recognize how examining carefully a problematic or incomplete strategy can be a very productive learning experience.

CLOSING
20 minutes

Materials
Handouts 9 & 10

Homework

10 minutes

Assigning Homework As you assign the homework, remind participants that this session was the first of two on the nature of the intellectual community in the classroom. Part of their homework assignment is to observe two teachers with respect to the intellectual community in their mathematics classrooms. Remind participants that in the first session of this course, they were asked to identify two teachers whom they could observe three times over the course and that guidelines and suggestions were offered (see p. 20). You may need to remind participants that they were to have done this in preparation for this assignment.

Explain to participants that the pre-observation and post-observation conferences are integral parts of the assignment and that they should plan to include them as part of their observations.

Spend a few minutes reviewing the Pre-observation Conference Questions (Handout 10). Explain to participants that the questions are designed to provide the observers with information about what the teacher intends to do in the lesson and how the lesson fits into the overall plan for the year. This information helps the observer make sense of classroom activities and events.

Bridging to Practice

10 minutes

Reflective Writing Remind participants that Bridging to Practice is designed to allow them to focus on the ideas in the course that seem particularly interesting for them and for their school community. Point out that it provides a framework for planning what they can do in order to move themselves and their schools along with these ideas.

Post the following questions, and invite participants to write about them:

- *Choose an idea that came up today that you found particularly interesting. What is your current thinking about this idea?*
- *Where is your school now with regard to this idea?*
- *What are one or two things that you, as instructional leader, will pursue to move yourself and/or your school along with this idea?*

Homework

The homework for this session includes both a reading and observations of two teachers in the participants' schools or districts.

A pre-observation conference is intended to help the observer make sense of what he or she will observe and to minimize the assumptions that might be made about what is happening in the classroom. A pre-observation conference can also create an environment of collaborative inquiry around the observation. When you review the pre-observation questions, you may need to point out that the pre-observation conference is not a planning session but a way for the observer to gather information about the lesson to be observed. The observer should use the pre-observation conference not to advise the teacher about what to do, but to find out from the teacher what to expect.

♦ **What topic will you and your students be working on in this lesson?** This question helps the observer know what area of mathematics will be covered in the lesson, such as fractions, two-digit subtraction, or area and perimeter.

♦ **What do you plan to do in this lesson?** This question gives the observer information on what learning activities the teacher intends for the lesson, such as how much small group, whole group, or individual work is planned for the lesson. The observer can also find out the genesis of the lesson: whether it comes from a particular curriculum or is a teacher-generated one. You may wish to suggest that participants request a copy of any investigation or problem beforehand and do the mathematics in advance, as they have done in this session.

♦ **What do you hope to accomplish in this lesson?** This question helps the observer know what the teacher's goals for the lesson are before the lesson begins. With this information, the observer is not distracted by trying to figure out the goals of the lesson during the observation.

♦ **What mathematical ideas are embedded in this lesson?** The purpose of this question is to move the teacher beyond the topics to be covered to an articulation of the fundamental mathematical concepts of the lesson. For example, in the videotaped clip presented in this session, the topic may have been patterns, but some of the deeper mathematical ideas of the lesson had to do with the properties of multiplication. As participants saw, some students discovered that they could not solve the problem for half the number of students and then double the answer because both the number of students *and* the number of valentines each student brings increase.

♦ **What have you and your students been working on prior to this lesson?** This question allows the observer to understand how the lesson follows from what the students were working on before and what knowledge and understanding students are bringing to the lesson. For example, if students have been exploring patterns in addition and subtraction, then a subsequent lesson that focuses on patterns in multiplication may raise some confusion because of the different ways in which additive and multiplicative patterns behave.

♦ **How does this lesson fit into your overall goals for the year?** The intent of this question is to help the observer gain insight into how the particular lesson fits into the teacher's long-term agenda for his or her students. For example, in the videotaped clip for this session, the teacher might have set as a goal that his or her students would learn to think systematically about the four operations and understand how they are interrelated. This particular lesson contributes to students' understanding of the multiplicative process.

♦ **Are there students who have special issues in the class?** This question gives the teacher an opportunity to share any special circumstances that influence his or her decision to structure or organize the lesson in particular ways. For example, there may be a student dealing with some personal issues, so the teacher will be less likely to call on that student. Or the teacher may have decided on particular groupings of students on the basis of personality factors or the nature of the ideas that students have been grappling with in the past.

Your Own Journal Writing

Within one or two days of teaching this class, after you have had the chance to unwind from the class and perhaps talk with a colleague or friend about how it went, set aside some time to write your own journal entry about the class. We encourage you to write about the ideas held by the participants in your group rather than your own actions and sense of how things worked out. In our experience, writing about what the participants are thinking about will be much more useful to you as you prepare for the next session. Some facilitators use the time when administrators are writing in their own journals to make preliminary notes for this journal writing.

Valentine Exchange Mathematics Problem

There are 24 students in Ms. Olivez's class. If each student gives a valentine to everyone else in the class, how many valentines will there be?

Discussion Questions for *Valentine Exchange*—Second Viewing

1. What is happening at the beginning of the class? What do you think is the initial problem that Ms. Olivez posed? What are some different ways of interpreting what she said?

2. How does the modeling and discussion at the beginning of the class seem to affect the strategies and approaches students select to try to solve the problem?

3. What are some of the mathematical complexities that the teacher and the students encounter as the activity progresses?

4. How does the intellectual community help bring to the surface these mathematical complexities? How might this become a constructive development?

Observation Guide

Students		
Focus Question	Conjectures	Evidence from Classroom
Math Content		
• What mathematical ideas are embedded in the lesson?		
• What makes this worthwhile mathematics?		
Learning		
• What kinds of mathematics sense-making are students doing?		
• What mathematical ideas seem to be confusing to students?		
• In what ways can you see that the students are developing their mathematical ideas over time?		
Intellectual Community		
• How are students showing respect for one another's ideas?		
• How do students use each other as resources as they make sense of mathematical ideas?		
• What evidence beyond raised hands do you have that students are engaged?		

Observation Guide

Teachers		
Evidence from Classroom	**Conjectures**	**Focus Question**
		Knowledge of Content
		• What does the teacher seem to understand about the mathematics?
		• What is the teacher's long-term mathematical agenda?
		• What does the teacher seem to understand about the development of children's ideas in this topic?
		Pedagogy
		• How does the teacher work with the sense the children are making?
		• How does the teacher work productively with students' confusion?
		• How does the teacher attend to all students?
		• How does the teacher adjust her teaching based on the ideas she hears from students?
		Facilitating Intellectual Community
		• How does the teacher support students in showing respect for one another's ideas?
		• How does the teacher set the tone so students see each other as resources for mathematical thinking?
		• What interventions does the teacher make to ensure that students' engagement has a focus on mathematical ideas?

Intellectual Community Observation Guide

Students		
Focus Question	Conjectures	Evidence from Classroom
How are students showing respect for one another's ideas?		
• What are students doing to show that they are listening to other students?		
• How willing are students to share their ideas even if they know that they aren't correct?		
• How attentive are students to one another's ideas?		
How do students use each other as resources as they make sense of mathematical ideas?		
• Are they building on each other's mathematical ideas?		
• Are they asking each other questions related to mathematical ideas?		
What evidence beyond raised hands do you have that students are engaged?		

Intellectual Community Observation Guide

Teachers		
Evidence from Classroom	Conjectures	Focus Question
		How does the teacher support students in showing respect for one another's ideas?
		• How does the teacher set and maintain norms of interaction for discourse?
		• How does the teacher support such practices as: • attentive listening • question-posing about mathematical ideas • provisional thinking
		How does the teacher set the tone for students to see each other as resources for mathematical thinking?
		• Does the teacher invite students' tentative thinking?
		• Does the teacher invite students to build on one another's ideas?
		What interventions does the teacher make to ensure that students' engagement has a focus on mathematical ideas?

Homework for Session 3

In the next session, we will be continuing our exploration of a mathematical intellectual community. We will begin to examine what it means to think mathematically and will also look more carefully at how students participate in intellectual communities. In preparation for this session, please do the following:

1. As a way of developing some images of classroom discourse, read *What's All This Talk About "Discourse"?* by Deborah Ball. (Note that this article will be discussed again in Session 6.)

2. Observe two teachers in your school or district teaching a mathematics lesson. Choose teachers who are not being evaluated for professional status and who are not encountering any problematic teaching issues. Use the attached Pre-observation Conference Questions to help you prepare for the observations. Use the Intellectual Community Observation Guide in the book of *Readings* for your observations.

3. Write up a short summary of each observation, addressing as many of the questions in the Intellectual Community Observation Guide as are relevant. Start with your overall impressions of this classroom. Then describe the intellectual community. Conclude with any questions or issues you are puzzled about.

Pre-observation Conference Questions

In order to help you make sense of what you will be seeing when you do your classroom observations, plan to meet with the teacher prior to the observation and ask the following questions:

1. What topic will you and your students be working on in this lesson?
2. What do you plan to do in this lesson? (e.g., the origin and structure of the lesson, and so on)
3. What do you hope to accomplish in this lesson?
4. What mathematical ideas are embedded in this lesson?
5. What have you and your students been working on prior to this lesson?
6. How does this lesson fit into your overall goals for the year?
7. Are there students who have special issues in the class?

SESSION
3

Linking Intellectual Community with Mathematical Inquiry

In this session, participants continue to develop their eye for the mathematics classroom as an intellectual community, concentrating on the inseparability of constructing an intellectual community and engaging in mathematical inquiry. In addition, participants consider how to recognize missing aspects of an intellectual community and decide what they would discuss with teachers at the post-observation conference. This last activity introduces the second strand of this course, *Rethinking their talk with teachers about mathematics, learning, and teaching,* and the notion of co-inquiry that underlies it.

Discerning the depth of an intellectual community from its surface features can be very challenging. For example, administrators may readily note that the teacher poses open-ended questions and has students work in small groups followed by a whole-class discussion in which students share their discoveries. The challenge, however, is to figure out whether there is a clear mathematical intent to these activities, whether they connect to a broader mathematical agenda, and whether the intellectual community is one in which students are engaged fully in the mathematical ideas embedded in the activities. In order to do this, administrators need to be able to connect the elements of an intellectual community with the mathematical content being explored. The activities in this session are structured to help participants learn more about how to do this.

Additionally in this session, participants are exposed to the algebraic concepts that underlie an elementary math problem. We address algebraic connections because algebra is considered a gatekeeper, but it is also the gateway for challenging and interesting courses in high school math and science. Teachers at the elementary level are expected to prepare students for the algebra they will take at the secondary level. Given this, administrators need to have a grasp of what constitutes good early preparation for algebra. They need to recognize that although algebra is about manipulating numbers and symbols, it also has a conceptual dimension that connects to important ideas in the elementary curriculum.

Overview for Session 3

OPENING page 84 10 minutes	**WELCOME AND INTRODUCTIONS** This is a time for announcements and introductions. It is also a time to provide an overview of this session and its connection to the overall course.
ACTIVITY 1 Homework Discussion page 86 30 minutes	**OBSERVING IN OUR CLASSROOMS: BUILDING INTELLECTUAL COMMUNITIES** Participants share with a small number of colleagues the classroom observations that they did for homework. They will have focused on the elements of the classroom that constitute an intellectual community.
ACTIVITY 2 Mathematics Exploration page 89 65 minutes	**ENGAGING IN MATHEMATICAL INQUIRY FOR OURSELVES** Participants engage in a rich, substantive exploration of the mathematics they will view in the videotaped clip during this session. In doing this exploration, participants gain experience with collaborative mathematical inquiry and with ways this type of community of inquiry can support students' developing algebraic sense making.
ACTIVITY 3 Video page 103 55 minutes	**MATHEMATICAL INQUIRY IN INTELLECTUAL COMMUNITY** Participants view and discuss a videotaped clip, *Products and Sums.* Participants investigate the nature of mathematical inquiry and the ways a teacher might more or less effectively draw out the mathematics in such an inquiry. They watch the clip twice, once with an eye on the mathematics of the investigation and a second time looking at the intellectual community of the classroom.
CLOSING page 114 20 minutes	**HOMEWORK** Participants read two articles, *Approaches to Solving a Fair-sharing Problem* (excerpt from the Fosnot and Dolk book–Reading 4) and *Magical Hopes: Manipulatives and the Reform of Math Education* by Deborah Ball (Reading 6). Prior to reading the first article, participants work on the same math problem that is the focus of the reading. **BRIDGING TO PRACTICE** Participants finish the session with journal writing that allows them to develop further some of their thinking from the session.

Big Ideas

Participants explore:

- the relationship between intellectual community and focused mathematical exploration
- the need for students to develop the skills of collaborative mathematical inquiry over time
- the teacher's role in shaping students' mathematical inquiry

Materials

- ☐ nametags
- ☐ flip charts or overheads
- ☐ markers
- ☐ agenda
- ☐ Handout 11, *Products and Sums* Pattern Exploration
- ☐ Videotaped Clip 3, *Products and Sums* (DVD #1; Program 17; first viewing, 0:45–17:17; second viewing, 9:40–17:17)
- ☐ Graph paper
- ☐ Square tiles
- ☐ Handout 12, Overall Observation Guide
- ☐ Handout 13, Intellectual Community Observation Guide
- ☐ Handout 14, Homework for Session 4
- ☐ Reading 4, "Approaches to Solving a Fair-sharing Problem" (excerpt from *Young Mathematicians at Work: Constructing Fractions, Decimals, and Percents* by Fosnot and Dolk)
- ☐ Reading 6, *Magical Hopes: Manipulatives and the Reform of Math Education,* by D. Ball

Preparation

- Review participants' writings, paying particular attention to the ways in which they have made sense of the notion of the intellectual community in the mathematics classroom. This will help you decide what to highlight or emphasize in this session, especially with respect to the discussion questions.
- Familiarize yourself with the Intellectual Community Observation Guide for Activity 1. Spend some time thinking through what evidence one might be looking for that would count as elements of an intellectual community. You may decide that you want to have a copy of the Intellectual Community Observation Guide on an overhead. Review the subsection "Orientation Toward Teachers when Observing" of the **Introduction** (p. 10).
- Reread the article *What's All This Talk About 'Discourse'?* by Deborah Ball for Activity 1. Pay attention to how the ideas about mathematical discourse presented by the author can help you construct grounded images of an intellectual community in a mathematics classroom.
- Explore the Products and Sums mathematical activity yourself for Activity 2. Try doing so before you read the **Facilitator Notes** for it. You will find it useful to pay particular attention to the strategies you use to find the

PREPARATION

relationships between the doubles and the squares. Also, note what properties of addition and multiplication and what relationships *between* addition and multiplication the activity clarifies. If you have time, you might want to ask a colleague or a friend who enjoys such mathematical inquiry to join you in exploring this problem. This would give you some personal experience with collaborative inquiry of mathematics.

- Think ahead of time about homogenous groupings for your participants. For this particular math activity, we have found the quality of the learning experience to be better when participants work in somewhat homogenous groupings regarding mathematical content knowledge. Mathematical exploration as a part of a learning community is the goal of this math activity. Because this is a first substantive mathematical exploration in this course, participants are more likely to be comfortable participating in this discovery together and less worried about the awkwardness of thinking differently or using different types of representations to solve the problem if they are in homogeneous groups. You will want a variety of levels of abstraction and use of algebraic language to come out in the discussion of the math, so ideally you will form at least four groups: one for participants with considerable experience and comfort with algebraic notation, one for those with moderate experience and comfort with algebraic notation, one for participants with comparatively less experience and comfort with algebraic notation but a very solid background with arithmetic descriptions of patterns and relationships, and one for those who consider themselves relatively new to standards-based mathematics but possibly experienced with standards-based classrooms in other disciplines.
- You will need to bring both graph paper and square tiles to the session. Be sure to plan ahead to have the tiles. You may have to borrow them from a teacher in your school or from another teacher in the district.
- After you have worked through the mathematics problem, preview the videotaped clip *Products and Sums* for Activity 3. As you watch the videotaped clip, pay attention to the ways the teacher structures the investigation, and think about what this might tell you about the mathematical relationships she would like her students to notice. Listen carefully to the discourse among the students as they work in their small groups. Notice, for instance, how they help each other construct strategies so that they can make increasingly subtle observations about the relationship between the sums and the products. You also want to think about what you would want to talk with this teacher about in a post-observation conference. Review the subsection "Setting a Climate for Viewing Videotapes of Teachers and Students at Work" (p. 9) in the *Watching Videotaped Clips of Teachers and Students at Work* section of the **Introduction.**

PREPARATION

- As we advocate in the **Introduction** ("Showing the Videotape Twice," p. 10), plan to leave enough time in the session for participants to view the videotaped clip two times. In this case the second viewing is of only a partial segment; still, you will want to allow the suggested amount of time. You may want to refer to the section *Viewing Videotaped Clips of Teachers and Students at Work* (p. 9) in the **Introduction** to remind yourself of the rationale that underlies this recommendation.

OPENING
10 minutes

> ***Materials***
> Name tags
> Agendas

Welcome and Introductions

Before starting the session, post the agenda in a place where everyone can see it.

5 minutes

Welcome After everyone has gotten their name tags and settled into their seats, welcome them to the third session of the course. You probably will not have any newcomers this time, but in case you do, you may want to do introductions again. If you do not have any newcomers, it would still be good to have everyone state their names and school districts. Encourage people to wear their name tags, especially if you have a large group.

5 minutes

Overview of Session Take about five minutes to go over the agenda for the session. Remember to link this session with the main ideas covered in the previous two sessions. This is especially important because Sessions 2 and 3 address similar issues and it will help participants to know what some of the differences are.

Remind participants that Sessions 2 and 3 focus on what constitutes an intellectual community in the mathematics classroom. Recognizing the nature of an intellectual community is a necessary precursor to reorienting one's focus when observing in classrooms. Tell participants that this session will highlight some of the features of mathematical inquiry, looking in particular at the ways in which a teacher can structure students' inquiry in an open-ended mathematical investigation.

Explain to participants that after they spend some time sharing the observations they did for homework, they will spend about an hour on a mathematical activity that is a modification of the activity in the videotaped clip they will see later in the session. After doing and discussing the mathematics, they will observe the clip twice. The first time, participants will view the entire video with an eye toward attending in a general way to the mathematics. The second time they will view only the middle segment, which shows students working together and sharing their findings with a focus on the identifying aspects of intellectual community that are present or absent.

Overview

By this session, you should have a good sense of your group. There will probably be no additional newcomers, and those who have decided that the course is not going to work for them will have let you know. This session is a good time to focus on what the issues and needs of the group will be over the next six sessions. As facilitator, you will need to help participants focus on the mathematics and connect the doing of mathematics with the cultivation of an intellectual community.

These **Facilitator Notes** are meant to give you an in-depth look at the mathematics in the videotaped episode and help you make those connections. Understanding the mathematics will prepare you to focus the discussions on the critical ideas about mathematics, learning, and teaching. Remember that the discussion questions that accompany activities are intended to prompt such discussions. Part of your work as facilitator is to use participant responses from the questions to go deeper into the ideas of the session.

ACTIVITY 1
Homework Discussion
30 minutes

Materials
Handout 12

Observing in Our Classrooms: Building an Intellectual Community

20 minutes
Small group

Sharing of Homework Observations Organize participants into groups of three or four, and give them five to ten minutes to share their observations. Encourage participants to sit with people who are not from their own district.

Talking about their unresolved questions will help participants move beyond describing a classroom on the basis of particular elements to reflecting on how the elements they observed contribute to student learning.

Explain to participants that they should take turns sharing their observations. Tell them to focus on two or three relevant features of student interaction or teacher facilitation that they found particularly interesting. Encourage them to share anything about which they may still be puzzling. For instance, they may have noticed that students were asking one another questions, but they may have been uncertain whether students were in fact building on one another's ideas. They also may have noticed the teacher encouraging tentative ideas but may not have been sure whether it contributed to the building of particular mathematical ideas.

10 minutes
Whole group

Discussing the Observation Guide After the participants have shared their individual observations, have them comment on how they found using the Observation Guide in classroom observations. Focus on how the Observation Guide assisted them in making sense of the intellectual community in the classroom they observed and what they found either challenging or puzzling in using it. Ask them to comment on the pre-observation conference questions. You might ask:

> *How did the information you obtained from asking the pre-observation conference questions contribute to your understanding of the lesson you observed?*

At the end of this discussion, tell participants that they will now engage in a substantive mathematical exploration. Explain that by doing the exploration, they will gain an appreciation for mathematical inquiry and will be prepared to view the videotaped clip that they will see later in the session.

Overview

In preparation for this session, participants read Deborah Ball's article, *What's All This Talk About "Discourse"?* as background before their first observation. Because the focus of the observation is on the intellectual community in the mathematics classroom, the article is intended to help participants focus on the features of classroom discourse that make a classroom an intellectual community. Although participants will not formally discuss this article today, you will want to look for ways to integrate some of the ideas into the discussion participants have about the observations they did.

Summary of the Article

A key purpose of the article is to consider how the NCTM's (1991) "Professional Standards for Teaching Mathematics" can be used as set of tools to construct productive conversations about teaching, but it also provides grounded images of what mathematical discourse entails. Ball examines the ways in which discourse is used to construct and exchange knowledge in classrooms and considers how teachers influence the norms of knowing and thinking by the discourse choices they make. One example she presents is the subtle difference between asking a student, "How do you know that?" and asking, "How did you get that?" As she notes, the first question often connotes that the answer is correct while the second question suggests that the answer may not be correct. She notes that when teachers hear answers as either right or wrong, and hear right answers as indicating understanding on the part of the student, they miss opportunities to gain insight into students' thinking. She follows this point with an example about a student who thought that $\frac{1}{6}$ was more than $\frac{2}{3}$. She illustrates how, having taken the time to explore what the student was thinking, she discovered that the student's understandings of fractions and subtraction were mixed together in a complicated way so that the student thought that $\frac{1}{6}$ meant that one had six parts and took away one and that $\frac{2}{3}$ meant that one had three parts and took away two.

Ball also presents an excerpt of a discussion that took place in her third grade classroom in which students explored what $\frac{3}{4}$ of 12 is. The classroom discussion is accompanied by Ball's own commentary in which she highlights the kinds of dilemmas and choices a teacher faces when listening carefully to the mathematical ideas with which students are working. Ball's comments point to the subtle ways in which teachers can shape the intellectual inquiry that takes place in the classroom by choosing carefully the questions they ask or comments they make.

Sharing of Classroom Observations

This activity is designed to give participants a chance to share their observations in the informal setting of a small group. Sharing publicly helps participants develop a more reflective stance toward classroom observation and a better understanding of how to engage in inquiry with colleagues. Sharing in the small group is also intended to provide a forum in which participants can share areas of confusion (e.g., items on the observation guide) or concerns that they may not want to raise with the whole group.

The small group discussions of the observations may also bring to the surface general issues that can be addressed with the whole group. As people share their observations and comments, in either small groups or the whole group, be listening for the common threads that can be topics of discussion with the whole group. Some threads may focus on the Observation Guide; participants may note that the Guide helped them recognize elements of an intellectual community that they had not been aware of before or determine what constituted evidence of an intellectual community.

Other threads might relate to ideas presented in the Ball article.

It is important to reiterate to participants that just because participants do not see evidence of an aspect that the Guide is asking about, they should not jump to the conclusion that it is not present in the classroom. It just may not be present in that particular lesson.

◆ **How did the information you obtained from asking the pre-observation conference questions contribute to your understanding of the lesson you observed?** Participants will have a range of responses to this question. A key understanding is that the information acquired from the pre-observation conference can help them bring a greater understanding to their classroom observations and not jump to conclusions about what they observe.

What participants learn from the pre-observation conference questions can feed productively into the ideas they discuss with teachers in the post–observation conference. For example, a teacher may explain in the pre-observation conference that her students will be working on different ways to make the number 10 and that students this age are at very different points in terms of their understanding of how to approach this problem—from using their fingers to making a chart that shows all the solutions. An administrator can bring this understanding of what students at this age typically do to the observation. Rather than questioning the teacher about the issue of using fingers to solve mathematics problems, the administrator and the teacher may wish to focus their post-observation discussion on other issues such as ways to help students extend or deepen their thinking.

ACTIVITY 2
Mathematics Exploration
65 minutes

Materials
Handout 11
Flip chart or overhead
Markers
Graph paper
Square tiles

Engaging in Mathematical Inquiry for Ourselves

10 minutes

Introducing the Investigation Explain to participants that they will spend about 30 minutes doing a mathematical exploration, a modification of the exploration that the students are doing in the videotaped clip. Let participants know that experiencing mathematical exploration as a part of a learning community is a central goal of the activity. Tell them that they have been grouped by experience level to provide an interesting and supportive learning environment and to encourage the full participation of all group members in the exploration. All groups will be doing meaningful work of mathematical sense-making and collaborative inquiry, regardless of the mathematical outcome of their discoveries or the way groups represent their findings. These findings will be shared with the whole group later.

You can note here that, in addition to helping participants gain an appreciation of mathematical inquiry and the role of the intellectual community in supporting it, doing the mathematics from the classroom episode beforehand gives them insight into the mathematical topic and potential of the lesson. Caution them that just doing the mathematics is not a substitute for talking with the teacher to find out what her intent and goals for the lesson are.

Organize participants into their groups. Pass out *Products and Sums* Pattern Exploration (Handout 11), and distribute graph paper and square tiles to each group. Tell participants that they will work by themselves for about five minutes on the first question and then, when they are ready, work together in their groups to address the remaining questions. Explain that they should concentrate on the second section of the worksheet, "Observing Growth Patterns," and move on to the third section only if they have exhausted their exploration of the second section.

25 minutes
Small group

Doing the Mathematics As people are working, walk around the room to listen in on the groups' conversations. Make sure that everyone understands the activity and is comfortable using graph paper or tiles to create visual representations of the numbers in each column. Participants may come up with a variety of representations; the most common is to create an array as the teacher in the video does. If a particular group seems stuck on this part, you might offer some prompts to encourage them to build arrays of the numbers. For example, you might ask:

> *How could you use the tiles to represent each of the doubled and squared numbers?*

This activity is intended to help participants uncover important characteristics of growth patterns of products and sums. Their explorations will go deeper than the general search for arithmetic relationships that takes place in the videotaped clip that follows.

After it seems clear that participants are comfortable with the activity, you will want to pay attention to the kinds of patterns they are noticing, the strategies they are using to find patterns, and the ways they are working with each other to explore the patterns and relationships. Notice, for instance, whether they are actually working together or working independently within the groups. Look especially for any instances in which they build on one another's observations to discover a systematic pattern.

Also pay attention to which participants, if any, are having difficulty with what it means to look for a systematic pattern or relationship. Some participants could potentially be stuck on relationships that look like systematic patterns but are not (e.g., noticing that $3 + 6 = 9$ and wondering whether the number plus its double always equals its square). Others might have noticed a pattern (e.g., the squares alternate between odd and even) but may not know what to do with this information or how it relates to a mathematical idea.

Take particular note of the insights about growth patterns of products and sums that different groups come to during this activity. During the discussion about the mathematics that follows, it will be important to draw out a variety of these ideas in order to help participants think about what central ideas and insights about addition and multiplication *students* might uncover when engaged in this activity. Having a number of different representations with varying degrees of abstractness will also help participants see the algebraic concepts that underlie this elementary math problem and prepare them to discuss the implications of a mathematical intellectual community.

30 minutes
Whole group

Discussing the Mathematics There are three distinct sections to the discussion, each building on the previous. First, participants share the patterns they discovered during their investigation. Next, they discuss the benefits of representing the information graphically or in a table in helping them think about, discover, and describe patterns. Last, participants reflect on their experiences with collaborative inquiry, comparing them with their math learning experiences as students, to gain insight into the powerful role of an intellectual community in learning mathematics.

Sharing Patterns After groups have had about twenty-five minutes of doing the mathematics, bring all of the groups together for a discussion. Begin by asking people to share some of the patterns they noticed. You might ask:

What patterns and generalizations did your group discover as you investigated these sets of numbers? Explain how your thinking developed to lead you to these conclusions.

During the discussion about the mathematics embedded in this activity, it will be important for you to draw on participants' ideas about addition and multiplication in order to help them think about the ideas and insights that students might uncover in such an activity. You may emphasize that generalized representation of concepts is key to algebraic thinking, a skill that students need to strengthen as they experience algebra more formally.

As a group shares a pattern, ask the members how they knew that the pattern would work for every situation. For example, one group might say that all of the doubled sums are even numbers and that they know that the generalization is valid because doubling is equivalent to multiplying by two, and any number multiplied by two is an even number. Another group might notice that the squares alternate between even and odd numbers, for which they might say that consecutive numbers alternate between even and odd and that two odd numbers multiplied by each other always results in an odd number.

After a group has shared patterns and the thinking behind them, ask another group to share patterns different from those already presented. You may want to call on a certain group if you noticed a distinctive and interesting pattern in their work. Continue until you have three or four different patterns. Leave the patterns posted during the next part of the discussion.

The Power of Representations and Generalizations Now probe for any insights into the relationship between addition and multiplication that the representations of the growth patterns suggest. You might ask:

> *What did you notice about the growth patterns when you represented them graphically? How can you describe the growth patterns you see?*

For groups that investigated formal algebraic expressions, you might ask:

> *What numerical or verbal statement did you come up with to describe these patterns? What process did you use in your search for a generalization that works?*

Have a flip chart and markers ready so that people can show what they discovered. Spend about 10 or 15 minutes sharing representations and the generalizations they generated.

If participants do not mention the connection between the graphic representations of the two growth patterns (e.g., the arrays drawn on graph paper) and the more formal algebraic expressions that can be used to describe these patterns, you might ask a series of questions to bring these ideas out yourself. You might start by asking:

> *Who can make a statement that describes in words the way a growth pattern works?*
>
> *How might this statement be written as an algebraic expression?*

The statement "The sum always increases by two" could be supported by thinking about "increases" indicating addition, or a "1" symbol that could lead to a list similar to the following: next sum = 8 + 2, then next sum = 10 + 2, and then next sum = 12 + 2. By noticing that "+ 2" is the same each time, participants can reach the generalization that the sum will equal the old number plus two, no matter what the old number was. Substituting n for the number will lead to the algebraic expression $n + 2$. Similarly, "The sum is always two times the starting number" can lead to "2 times the number," or $2n$. See the **Facilitator Notes** for an expanded discussion of the algebraic equations that underlie the products and sums problem.

When you discuss the algebraic expressions, be sure to be clear with participants that fourth graders who would not yet be familiar with such formal mathematical generalizations and notation and would not be expected to write algebraic expressions. They would, however, be expected to identify and discuss the patterns in less formal ways.

Experiencing an Intellectual Community After participants have shared the patterns and relationships they have discovered, shift the discussion to their experiences doing the activity. Focus first on the experience of engaging in mathematical inquiry. Some of the questions you can ask the group include the following:

What did you find challenging or engaging about looking for patterns and relationships in a group setting?

How was this investigation different from the process you experienced when you were a middle or high school student? How was it similar?

You will also want to draw participants' attention to the qualities of collaboration. Examples of questions you might ask are the following:

In what ways did you serve as resources to one another as you made sense of mathematical ideas?

What examples can you cite of building on one another's ideas?

Overview

This extended investigation builds on the purposes of doing mathematics stated in Session 2. That is, it allows participants to focus better on the videotaped classroom episode by not having to figure out the problem while viewing it. It also helps participants gain a deeper appreciation of the nature of mathematical inquiry by engaging in such inquiry themselves. By doing the mathematics ahead of time, participants can appreciate the mathematical potential of the lesson. Throughout this course, participants do the mathematical problem themselves before observing a videotaped clip so that they can appreciate the importance of content knowledge for supervisors carrying out classroom observations. In this particular instance, the mathematical exploration has an additional critical goal—that of genuine discovery and deeper understanding of both the mathematics involved and the dynamics and process of group learning through collaborative inquiry or investigation.

When participants engage in this collaborative mathematical investigation, they will be exploring a rich mathematical terrain. The assigned problem has significant potential. It is appropriate and interesting for fourth grade students to explore these patterns (doubling and squaring) using arithmetic. The problem also holds potential for the emergence of algebraic thinking (e.g., generalizing, developing rules to describe change, abstracting) and possibilities for connection to the language of formal algebraic notation. It is important to note that problems that are not obviously algebraic in the use of formal expressions can be *conceptually* algebraic. That is, participants may come to realize that in working through this problem, elementary students grapple informally and intuitively with algebraic ideas such as linear and quadratic patterns of growth that they will later examine explicitly in the context of formal algebra.

Learning through Mathematical Inquiry

Learning systematic ways of investigating the nature of numbers and operations is a critical skill that students must develop as part of problem solving. Through such investigations, students not only learn how to approach and make manageable complex mathematical problems, but they also discover important aspects of the base 10 number system and the four operations—knowledge they will need in order to hone their computational skills. Engaging in such systematic investigations helps students prepare for more advanced mathematics, such as algebra and calculus, that involve working with mathematical abstractions and generalizations.

In this investigation of the similarities and differences between the growth patterns of a number doubled and a number squared, participants will also have an opportunity to use their own problem-solving skills. For participants to begin to understand the interconnected nature of intellectual community and the learning of mathematical content, they need to engage in substantive exploration themselves.

For this particular activity, we have found the quality of the learning experience to be better when participants work in somewhat homogenous groupings regarding mathematical content knowledge and comfort level. Mathematical exploration as a part of a learning community is the goal of this mathematics activity. The discoveries made by different groups may or may not be similar, but each participant will be doing the meaningful work of mathematical sense making and learning about the process of working collaboratively.

Because most adults learned mathematics through a very different process, they can best begin to appreciate the power of the standard-based practice of participation in an intellectual community by gaining firsthand experience in such a community themselves.

Understanding the Mathematics

The problem of comparing sums and products of positive whole numbers contains substantially more mathematical richness than might be evident at first. Many of us are unaccustomed to thinking of elementary mathematics as having important elements of algebraic ideas and concepts, so it may be surprising to discover that exploration of sets of sums or sets of products can lead to generalizations that may be written algebraically. The algebra of the growth patterns involved in doubling and squaring provides a fertile mathematical terrain for participants to explore, particularly when connecting formal algebraic expressions such as $y = 2x$ and $y = x^2$ to the same patterns described arithmetically. There is added mathematical potential in comparing the relationship between the sum and the product of a number, as the teacher in the classroom episode suggests that her students do.

The growth patterns of doubling and squaring can be investigated arithmetically or with varying degrees of abstraction, including formal algebraic notation and graphing them as functions.

The first task for participants in this activity is to make a table showing sums and products of positive whole numbers from 1 to 10, similar to the one shown in Table 1.

Number	Sum	Product
1	2	1
2	4	4
3	6	9
4	8	16
5	10	25
6	12	36
7	14	49
8	16	64
9	18	81
10	20	100

Table 1

♦ **Adding a Number to Itself** By examining the number and sum columns and displaying this information graphically, participants (or students in elementary classrooms) can see doubling as a phenomenon, or, in more formal, algebraic terms, as a function, a rule that assigns a unique output (in this case either the sum or product) to each given input (in this case, the number). Functions are powerful mathematical tools because of their generalness. By making generalizations, the learner can concentrate on the big picture, or the pattern of behavior for a group of numbers. Additionally, using an explicit function such as $y = 2x$ to describe a pattern of numeric behavior allows the learner to compute the outputs for specific inputs very easily. Organizing and representing numbers in charts and graphs helps the learner notice characteristics such as the steadiness of the growth of sums compared with the ever-increasing growth of products. Charts make certain patterns more obvious, such as that all the sums are even, but graphs illuminate other trends such as that although the results are increasingly positive in either case, a straight line fits one pattern (doubling) and another type of curve fits the other (squaring). We discuss the case of doubling first.

♦ **Doubling** Even without organizing the numbers and their sums in this way, everyone has an idea of what doubling is. One common way to think of doubling is in the series 1, 2, 4, 8, 16, and so on. Similarly, we know that 5 doubled is 10, or that $1\frac{1}{3}$ cups doubled is $2\frac{2}{3}$ cups. All of these facts, however, stem from thinking of doubling in solitary instances. Even though doubling is a simple idea, the power of the activity comes from the insights to be gained when doubling is viewed as a *pattern.* This way of looking at numbers leads to generalizations, such as that successive sums increase by 2, that all of the doubles are even, or that a number added to itself is the same as a number multiplied by two. These generalizations are an important component of algebraic thinking.

One feature of the pattern of doubling can be observed by examining the raw numbers in the Number and Sum columns in Table 1. Notice that as one moves down the Number column, the quantity increases by one each time. As one moves down the Sum column, the quantity increases by two; in essence, skip-counting by two (e.g., 10, 12, 14, 16). This relationship allows us to predict the next number in the sum column by adding 2 to the previous number. We can also find the value in the Sum column for any number by multiplying that number by 2. This holds because the sum is the number added to itself, or the number taken *two times.* Using formal algebra, we can write this relationship as $y = 2x$, where y represents the double of x, or the sum of x plus x.

One helpful step in looking for patterns and generalizations can be making a chart, as in Table 1; another is making a graph. Below are two graphs displaying the original number and its double. The first consists of counting units and shading them (Figure 1); the second uses the (x, y) coordinate grid, a common tool of algebra (Figure 2). Both graphical representations compare the two quantities for several numbers, and both yield insight into doubling that cannot be gleaned from cases in isolation, such as the exact predictability of the incremental increase, by two with each step, or the precise linear nature of the pattern of growth.

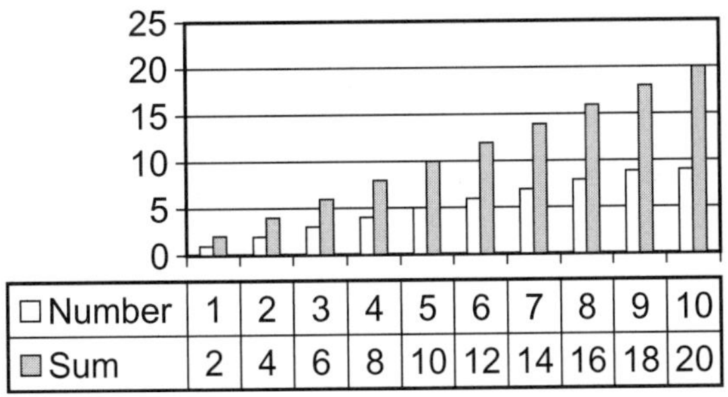

□ Number	1	2	3	4	5	6	7	8	9	10
▨ Sum	2	4	6	8	10	12	14	16	18	20

Figure 1

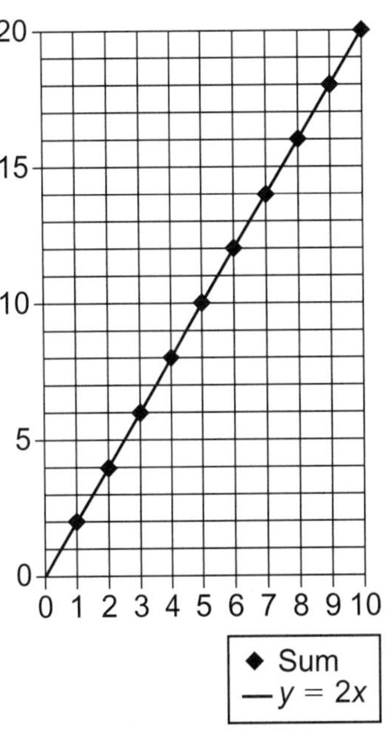

Figure 2

In the second representation, a line may be drawn through the dots that represent actual information from the table. This is an example of the usefulness of this type of graphing and of the power of algebraic representations—it can be inferred that doubling is regular and continuous and that it holds true for fractional values as well as for whole numbers. This is in fact the case. Not all patterns or functions can be extended in this fashion, of course, because not all patterns are continuous and without a limiting end point, but the act of displaying a relationship graphically remains an important part of understanding it.

◆ **Multiplying a Number by Itself (Squaring)** The result of multiplying a number by itself is often called a square number, and in fact, all such numbers can be represented on a grid as squares. Notice below successive square values represented in this fashion.

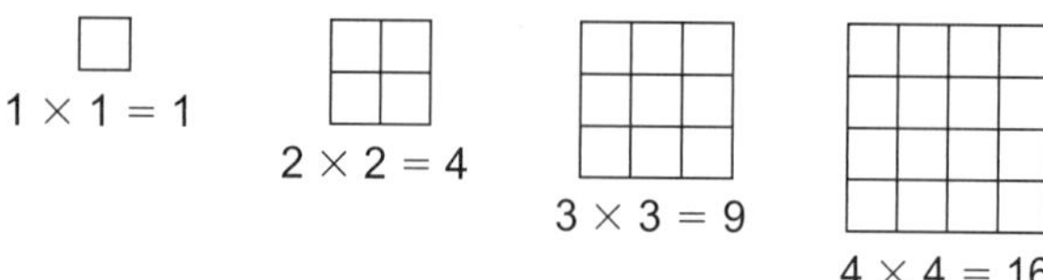

Figure 3

The pattern of squaring, or multiplying a number by itself, can also be represented graphically. Note that the difference between successive squares changes much more rapidly than the difference between successive sums. Consider the following graphs (Figures 4 and 5), and compare them with the similar graphs for a number and its double (Figures 1 and 2).

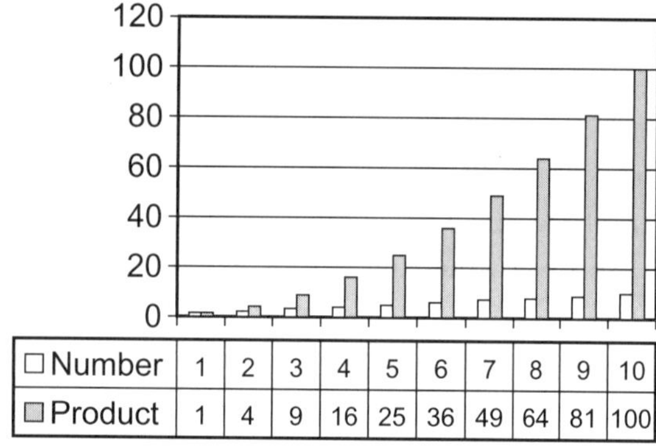

☐ Number	1	2	3	4	5	6	7	8	9	10
▩ Product	1	4	9	16	25	36	49	64	81	100

Figure 4

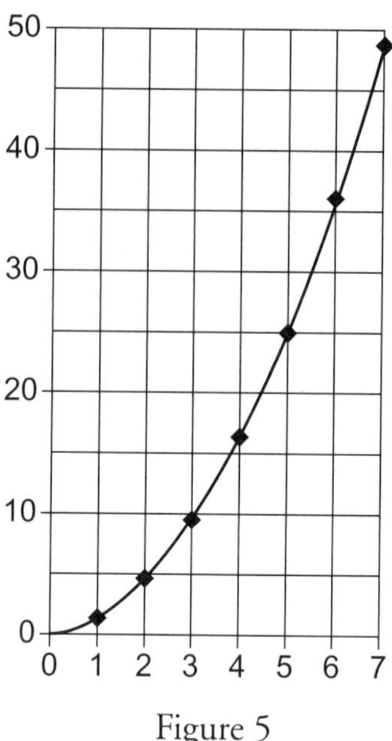

Figure 5

Several interesting facts can be noted from this way of representing quantities. For example, successive squares increase by successive odd integers. That is, the product of 2 × 2 is **3** larger than the product of 1 × 1, the product of 3 × 3 is **5** larger than the product of 2 × 2, and so on. Also, square quantities alternate between having odd and even values; 1, 4, 9, 16, 25, and so on. (This occurs because an even number squared is even, but an odd number squared is odd.) Still more interesting is the geometric comparison of one square to the next. As you can see in Figure 6, the smaller square fits into a corner of the larger one, with some area remaining.

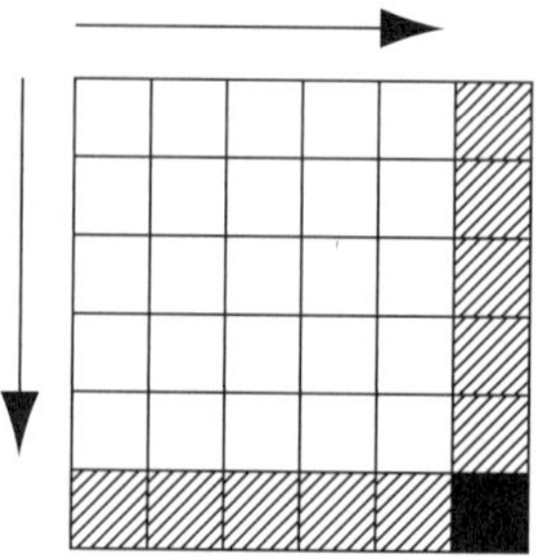

Figure 6

The remaining area can be accounted for by expanding the smaller square in one direction, for example, to the right by one unit, then in a perpendicular direction, for example, down by one unit. Then only one more unit square needs to be added to the corner, and the result is the same size as the larger square.

Using formal algebra, we can describe this increase as n^2 (the original square) $+ 2n$ (two times the number, or expansion by the amount of the number in two directions) $+ 1$ (to account for the last corner) $= (n + 1)^2$ (or the square that belongs to the next larger number). If we remove the explanations and leave the equation alone, we have the following quadratic equation:

$$n^2 + n + n + 1 = n^2 + 2n + 1 = (n + 1)^2$$

Another way to account for the increase is by expanding the smaller square one unit in a direction, for example, to the right, and then expanding the *entire* new rectangle in a perpendicular direction, for example, down by one unit. Because the rectangle used in the downward expansion is one unit larger than the original square, we must account for that by calling the second addition $n + 1$. See below for a graphic representation.

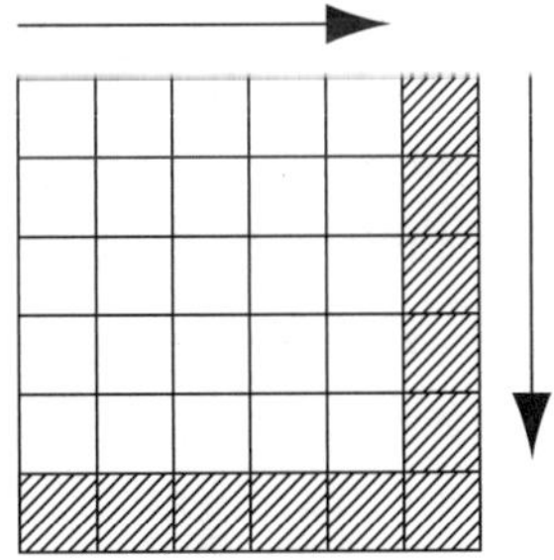

Figure 7

This method of accounting for the difference between successive squares yields the expression n^2 (the original square) $+ n$ (the first expansion) $+ (n + 1)$ (the second expansion). There is no need to add the last single unit to fill the larger square completely because the second expansion already accounts for it. Once again, the equation can be simplified from

$$n^2 + n + (n + 1) = (n + 1)^2$$

to

$$n^2 + 2n + 1 = (n + 1)^2$$

♦ **The relationship between the double and square of a number.** This is the aspect of sums and products that the teacher in this session's videotaped clip instructed her students to investigate. For this relationship, study the set of arrays in Figure 8.

Number	Sum	Product
1	2	1
2	4	4
3	6	9
4	8	16
5	10	25
6	12	36
7	14	49
8	16	64
9	18	81
10	20	100

Figure 8

Notice that if you double a number and multiply that by half the number, the result is the square of the original number. To see the pattern in the arrays, consider the example of the sixth row. Half of six is three, and if you take the sum array and stack three of them side by side, they form the square seen in the product column. This is also easy to see in the other even columns, and it can be seen in odd-numbered columns with slightly more effort (for example, half of five is $2\frac{1}{2}$, so the square should be $2\frac{1}{2}$ times the double of 5). The double of 5 is an array of 10 squares. To multiply using the graphics of Figure 8 as a tool, find the row for the number 5, and then find the sum array and chunk together two of them, plus another half, split along the long axis. This results in the array in the products column. This relationship can be written algebraically as

$$2n \times \frac{n}{2} = n^2$$

You can also see how this would work in a graphic representation as shown in Figure 9. If you take an n by n square, cut it in half, and place one half on top of the other half, the dimensions of the resulting rectangle will be half of n by two times n.

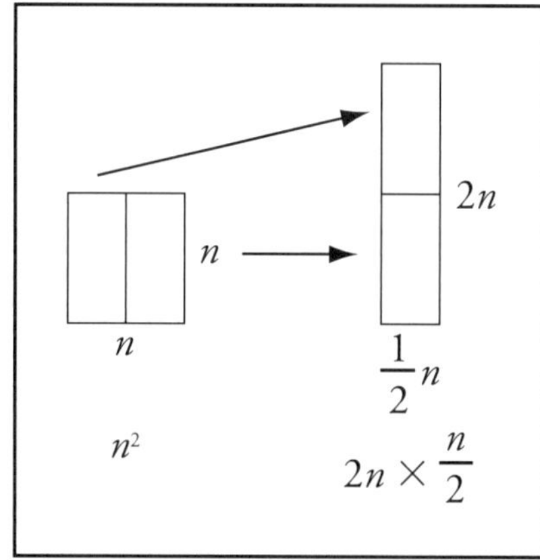

Figure 9

Doing the Mathematics

The activity is constructed to be open-ended enough for participants to discover a range of patterns and relationships and yet structured enough for them to discover (at least potentially) several key patterns, particularly the difference between the growth pattern of doubles and squares. Some participants will gain a deeper understanding of the patterns without noticing the formal algebra underlying them, others will probably discover algebraic generalizations to describe the patterns, and still others may actually use algebraic expressions in their descriptions of their findings.

Participants are asked first to fill in the chart by themselves and to note any patterns they would like to explore further. The act of filling in the chart is meant to help them start thinking mathematically about the problem. As they fill in the chart, they will likely notice a few differences between the two columns.

Specifically, they will undoubtedly notice that the sum column increases consistently by 2 (i.e. 2, 4, 6, 8, 10, and so on) while the products column increases more quickly (i.e. 1, 4, 9, 16, 25, and so on). Some participants may notice that the products alternate between even and odd values and that the amount of the increase in the products column is represented by consecutive odd integers (i.e., 3, 5, 7, 9, and so on; see Table 1).

The second part of the activity, during which participants work with one another, focuses on the growth patterns, or the rate at which the columns increase in value, in the two columns. As noted above, they increase very differently: sums increase steadily by 2 each time, and products increase by greater and greater amounts each time. In this part of the activity, participants explore in a more systematic way *how* the numbers increase in value and link their findings back to differences between addition and multiplication.

Participants use graph paper or tiles to create visual representations of the two columns. Some participants may represent these quantities as single stacks of tiles or columns one unit wide drawn on graph paper, which will look like Figures 1 and 4.

Others may use the graph paper in a familiar (x, y) coordinate grid and graph products and sums as the functions $y = 2x$ and $y = x^2$, which will look like Figures 2 and 5.

The representations that participants are most likely to come up with, however, are those that Ms. Doolittle, the teacher in the videotaped clip, provides to her students. She constructs arrays for the numbers in each column using graph paper, as shown in Figure 8.

By constructing the arrays themselves, participants will more likely notice several characteristics of the growth patterns of the columns. In the case of the sums, each sum is two more than the previous sum. This rate of increase stays constant. The resulting array is a series of rectangles that increase in size by two. Some participants may be surprised when they see that the value in the sum column is twice the original number. This discovery is one example of how different exploratory mathematics investigations are from the traditional instructional methods that were in practice when many adults learned elementary mathematics. Adults generally know that 2 times 4 is 8, 2 times 5 is 10, and so on, but many of us learned these facts without a any sense of what doubling looks like as a *pattern* when applied to consecutive numbers.

Participants generally notice that all of the square numbers can, in fact, be arrayed as squares, which explains why multiplying a number by itself is known as squaring the number. As they construct the squares, participants might notice that the *difference* between number of tiles or squares for each consecutive pairs of square numbers *increases* by two.

That is, 2^2 uses **3** tiles more than 1^2, 3^2 uses **5** more tiles than 2^2, 4^2 uses **7** more tiles than 3^2, and so forth.

There are various ways to express this increase in difference. Some people might write verbal statements such as the following:

> The difference between each pair of consecutive square numbers will increase by two.
>
> The differences between squares are sequentially increasing odd numbers.
>
> The square of a number will be bigger than the previous square by two times the previous number plus 1.
>
> The square of a number is bigger than the previous square by that number plus the next number.

Some people might write a numeric or algebraic statement that captures this growth pattern. Two such versions are as follows:

$$(n + 1)^2 = n^2 + n + n + 1 = n^2 + 2n + 1$$

$$(n + 1)^2 = n^2 + n + (n + 1)$$

These two statements are algebraically equivalent but emphasize different ways of accounting for the increase. You can see this in the following two diagrams illustrating the increase from 5^2 to 6^2: (See **Understanding the Mathematics** for a fuller explanation of this.)

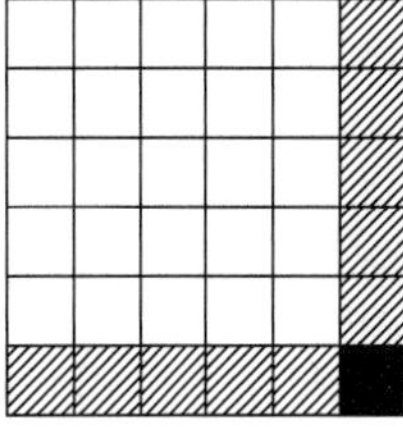

$$(n + 1)^2 = n^2 + n + n + 1 = n^2 + 2n + 1$$

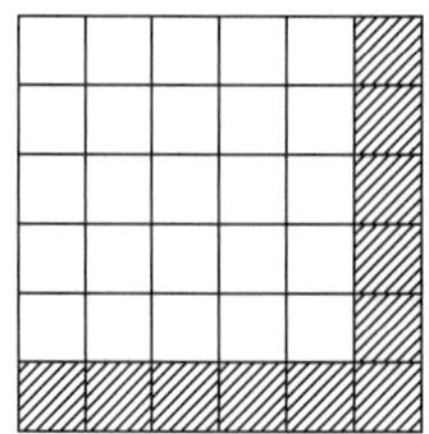

$$(n + 1)^2 = n^2 + n + (n + 1)$$

When you discuss the algebraic representations, emphasize that such representations would not be expected of fourth graders, who would not yet be familiar with such formal mathematical generalizations or notations, although they may be familiar with discussing patterns more generally.

Participants who use the (x, y) coordinate graph may note that the two functions illustrate linear behavior (or constant rate of growth) in the case of the sums ($y = 2x$) and quadratic behavior (or linear rate of growth) in the case of the products or squares ($y = x^2$).

They may note that the slope of the line $y = 2x$ is 2 and compare this with the standard slope-intercept form for linear equations, $y = mx + b$, where m is defined as the slope (or change in y divided by change in x between two points) and b is defined as the y-intercept (or the point where the line crosses the y axis, where $x = 0$).

For the line graph of the products, participants may note that they have drawn half of a parabola and realize that the numbers they squared did not contain any negative values (the domain is the set of positive integers).

The third section of the worksheet is an extra piece that is included, in part, because of the way in which Ms. Doolittle initially sets up the problem. She asks whether there is a relationship between a number added to itself and a number multiplied by itself. One relationship that some participants might notice is that if you double a number and multiply it by half the number, the result is the square of the original number. This can be written algebraically as follows:

$$2n \times \frac{n}{2} = n^2$$

As is explained in **Understanding the Mathematics,** if you take a square, divide it in half and put one half on top of the other half, the dimensions of the resulting rectangle will be half of n by two times n. (See Figure 9.)

Discussing the Mathematics
Sharing Patterns

• What patterns and generalizations did your group discover as you investigated these sets of numbers? Explain how your thinking developed to lead you to these conclusions. As mentioned above, participants may uncover differences in the growth patterns of the sums and the products of consecutive numbers. It is important that participants notice the different growth patterns but also understand *why* these differences exist or what is happening mathematically to create those differences.

Participants may also notice arithmetic relationships between number facts that are independent of the relationships between products and sums, such as the ones articulated by students in the videotaped clip (see **Facilitator Notes** for Activity 3 for further explication). These are intriguing to uncover, but it is worth noting that some of these relationships are mathematically significant and others are the result of arithmetic coincidence. Again, it is worth asking the question *why* about these patterns to uncover the mathematical relationship that underlies an observation. Such questions can also reveal the absence of a mathematical relationship if what seemed to be a pattern is actually mere coincidence.

The Power of Representations and Generalizations

♦ **What did you notice about the growth patterns when you represented them graphically? How can you describe the growth patterns you see?** Visual representations help uncover important relationships between addition and multiplication. Too often the patterns that are at the heart of mathematics have been made inaccessible because learners are introduced to equations without looking at what is occurring mathematically. Students have often been taught to memorize formulas and equations rather than make sense of what is happening in growth patterns.

In this activity participants are encouraged to use visual representations of growth patterns to connect to the logic of the mathematical relationships that can be observed in the columns of sums and products. In so doing (even in the absence of formal equations), participants engage in algebraic thinking that helps them uncover important truths about how the operations of addition and multiplication work.

It is fairly easy to visualize what is happening when one adds a number to itself: in the arrays we used earlier, the height of the column remains the same, but the width increases because of the addition of another column of tiles. What happens, though, when we multiply a number by itself? In this case the array increases horizontally *so that it is exactly as wide as it is high—and both dimensions of the array are that of the original number.* One interesting point to reflect on is that the array of 16 is made up of 4 columns of 4 tiles. When we move from the square number 9 (where n is 3) to the square number 16, we do not add 8 tiles (i.e., a row of 4 tiles and a column of 4 tiles, as some might expect). Nor do we add 6 tiles (a row of 3 tiles and a column of 3 tiles). Rather, we add 7 tiles ($2n + 1$ (because n is assumed to be 3 in this case)). Why is this? One way to explain this is that each of the 3 columns of 3 tiles from the 9 square requires another tile to make them columns of 4, so 3 tiles are added, and then a fourth column of 4 tiles needs to be added as well. (See Figures 6 and 7.)

♦ **For groups that investigated formal algebraic expressions, what numerical or verbal statement did you come up with to describe these patterns? What process did you use in your search for a generalization that works?** One interesting observation to explore is the difference between the patterns for addition and multiplication. Instead of growing by a constant number as do the numbers in the sum column, the numbers in the product column increase in a predictable and linear way that can be represented by the expression ($2n + 1$). (See explanation in **Understanding the Mathematics** in this activity.)

Experiencing an Intellectual Community

♦ **What did you find challenging or engaging about looking for patterns and relationships in a group setting?** Some participants may have found that working in groups was distracting and that they could not concentrate well because there were too many ideas being shared. This should be respected. For many of us, this can be a downside to working collaboratively. If this should come up, you might mention that many people find working in groups confusing or distracting at some times but helpful at other times. The key is to find a productive balance. You can also reflect on when you did the mathematics activity yourself. Were there times that you liked working by yourself and other times when you really would have appreciated a collaborator?

♦ **How was this investigation different from the process you experienced when you were a middle or high school student? How was it similar?** An important point to bring out in the discussion is how the algebra that emerged in this session was about more than manipulating numbers and symbols. The formulating and representing of generalizations that participants did grew out of the sense-making they engaged in as they explored the mathematics.

Participants may note that the meaning of the patterns at the heart of mathematics have been made inaccessible to many learners who were introduced to equations without the opportunity to sort out what is occurring mathematically. Rather than making sense of what is happening within growth patterns, generations of students have been taught to memorize formulas and equations. In this activity participants have used visual representations of growth patterns to connect to the logic of mathematical relationships. In so doing, whether they represented their findings as an equation or not, they have engaged in algebraic thinking that uncovers important truths about the operations of addition and multiplication.

It is important to point out to participants that if elementary students do not develop their ability to make sense of mathematical concepts, they run the risk of having a weak foundation on which to build their understanding of algebra when they encounter it more formally. Participants need to understand that this is what is at stake if teachers do not support this kind of sense-making in their classes.

- **In what ways were you resources to one another as you made sense of mathematical ideas? What were examples of building on one another's ideas?** Be prepared for a number of different answers to this discussion question. Certainly, it is hoped that participants would have found that working with others helped them see patterns and relationships that they would not have found otherwise or helped them understand more fully the pattern that they were noticing because they had to help their partner or other group members understand what they were noticing. In general, working in a collaborative inquiry setting helps people get further in their understanding than they would working alone.

As participants respond, ask for specific examples of how sharing one's evolving thinking either allowed them to get a fresh perspective on the ideas by virtue of articulating them aloud or provided a stepping-stone for another participant to build on as ideas were collaboratively constructed.

ACTIVITY 3
Video
55 minutes

> ***Materials***
> Videotaped Clip 3
> Handout 13
> Flip chart or overhead

Mathematical Inquiry in Intellectual Communities

25 minutes
Whole group

Discussing the Videotaped Clip, *Products and Sums*
Tell participants that they will watch a video of a classroom activity in which students explore the same mathematics that they just did.

> The value of this activity lies not in critiquing the teacher but in thinking about how to work *with* the teacher in a mutual pursuit of understanding student learning and building an intellectual community of mathematical inquiry.

Explain again to participants that the central purpose of viewing and discussing videotaped clips in this course is not to critique the teaching depicted there. Although participants might be tempted to do this, they simply cannot know enough to make valid judgments about the teaching. Teaching is complex, and from the perspective of an outside observer, there are always other moves that might have been made. An outside observer who sees only a short excerpt of a lesson, such as what participants see in these videotaped clips, will not know enough about the background of the students, the context of the lesson, or the teacher's thinking to make valid judgments about whether the teacher's decisions were good ones.

Give participants the following background information about the classroom:

- This is a fourth grade class.
- Students explore patterns and relationships in numbers that are doubled and squared.
- The 15-minute videotaped clip has been edited to show portions of the introduction, small group work, and whole group discussion of the investigation. Occasionally one also hears the teacher, Ms. Doolittle, in a voice-over reflection.
- Participants will watch the videotape twice. The first time (0:45–17:17) they will watch the entire tape with an eye to what is happening mathematically. The second time (9:40–17:17) they will review a shorter segment and focus on the intellectual community.

After the first viewing of the videotaped clip, spend about 15 minutes discussing it. Ask participants for their general impressions. You might ask:

What did you think of this class?

Asking this will open the discussion as well as help you gauge how the group perceived the lesson. Listen to the comments participants make and the questions they raise. Did they find the investigation compelling? Was it difficult to follow? Did they find the validity of the students' patterns difficult to assess? Use this information to shape the discussion of the mathematical ideas at play.

Exploring actual classroom data in terms of what one sees and is curious about is an important exercise in supervisory practice. This activity helps participants refine their observational skills.

Next, initiate the discussion about what is happening mathematically. At this point you might ask:

What mathematical ideas do you think students are grappling with in this investigation?

If participants respond with general answers such as "developing the ability to find patterns" or "learning how to problem solve," you might want to probe a bit. Ask, for instance, why they think finding patterns is mathematically significant, or how this investigation contributes to understanding both the relationships between sums and products and the power and role of prediction in mathematical reasoning and problem solving. Remind them of the rich mathematical terrain of growth patterns of addition and multiplication, and help them consider how students were supported in uncovering these ideas in this particular activity.

If participants express disappointment that students did not fully explore connections to algebraic concepts, remind them that the focus in this activity is on the mathematics the students engaged in and the ways students worked together to grow ideas.

Next, focus attention on the students' mathematical inquiry. Be sure to attend to how individual students are making sense of the mathematics as well as how they work together to build on one another's ideas. You can use a question like the following to prompt the discussion:

How are students making sense of what it means to search for patterns and relationships?

Within the context of the mathematical richness of the problem, these data form the base for working collaboratively with the teacher to enhance student learning.

10 minutes

Second Viewing of the Videotaped Clip Distribute a clean copy of the Intellectual Community Observation Guide. Explain that for the second viewing, participants will view only a portion of the video. This portion shows the students working together and reporting their findings and follows the full videotaped clip on the same tape. Emphasize that participants should be thinking about how the intellectual community contributes to mathematical inquiry.

Give participants a minute to review the contents of the Intellectual Community Observation Guide. Tell them that they are to attend to these questions as they view the segment of the videotaped clip a second time. Remind them that their focus is on identifying aspects of an intellectual community and the collaborative dynamic in the classroom that helps create the learning environment.

Remind participants that they should not jump to conclusions about the nature of the intellectual community in the classroom if they see no evidence of a particular item on the Guide. Rather, they should think of the Guide as providing prompts for questions that they might want to talk about with the teacher.

After showing the videotape, give participants a few minutes to jot down their observations about the nature of the intellectual community in the classroom.

20 minutes

Whole group

Discussing the Intellectual Community Spend about 10 minutes discussing aspects of the intellectual community in this classroom. Have participants talk about which aspects of an intellectual community they found to be present and which aspects seemed to be absent. You can ask the following questions:

How are students showing respect for one another's emerging ideas about arithmetic and multiplicative patterns?

How does Ms. Doolittle support students in showing respect for one another's ideas?

How do the students use one another as resources to build on their own understanding?

When participants seem to have developed a sense of the nature of the intellectual community in this classroom, explain that they will now shift their thinking from making sense of what is happening in the classroom to considering what they would talk about with the teacher in a post-observation conference. Emphasize that they should consider questions about which they are genuinely curious,

> Participants may express frustration that the classroom episode shows respectful listening and cooperative behavior but not intellectual depth of engagement with the mathematics. Encourage participants to think about effective support of teacher development that can address the noted absence.

such as students' engagement with the mathematical inquiry, rather than questions that seem to focus on assessing the teacher's pedagogical decisions and moves. Remind participants that their primary purpose in observing this classroom is to gain a better sense of how an intellectual community can support mathematical inquiry and not to assess the classroom. You can open the discussion by asking:

> *What questions about the student thinking that took place in this episode might be fruitful to discuss with this teacher?*

Facilitating an intellectual community in a classroom is hard work, and teachers need support and professional development opportunities in order to grow in their practice. Participants can consider ways to engage with the teachers as co-collaborators in the quest for deep student engagement with mathematical ideas.

Write down the questions that participants offer on a flip chart. After you have written about four to six questions, step back and ask whether there are any patterns to what people are curious about. Notice, for instance, whether the questions that participants offer are substantive or more process-oriented. Are they able to suspend an evaluative stance to ask more inquiring questions?

Synopsis of the Videotaped Clip

This videotaped clip depicts a fourth grade class in which students are exploring growth patterns with doubles and squares. The teacher, Ms. Doolittle, starts by inviting a student to write on the board the expressions 4, 4 + 4, and 4 × 4, and asking the class how they might investigate any relationships among these expressions. She notes in a voice-over that she thought it was appropriate for students to investigate relationships between addition and multiplication after a unit of study on multiplication and patterns.

She engages the students in a whole-group discussion about how they might investigate any relationships. One student suggests trying another four and then offers some preliminary observations. Another student suggests that they try out other numbers. Ms. Doolittle prompts the students to consider how they might investigate patterns on the basis of what they have done before. One student notes that they made models and gave examples. At this point Ms. Doolittle models the process of creating a three-column table: one column for the number, one for the sum, and one for the product. She shows them how to draw arrays for each value in each column. She then has the students work in small groups to investigate one number, its double, and its square.

After the students have had some time working on this, Ms. Doolittle brings them together for a whole group discussion of the models the groups posted at the front of the room. The students discuss their observations for a while, and then Ms. Doolittle asks them how they might compare the different models. One student proposes that they could put the numbers in order from the lowest to the highest. At this point, Ms. Doolittle provides them with a table that she made ahead of time of the numbers 1 through 9, with their respective doubles and squares and accompanying arrays for each of the values. She instructs students to work in small groups to explore the relationships between the sums and the products.

At this point, the video picks up on different groups of students exploring patterns. Some of the observations students make include:

> The numbers increase by one, the sums "go by 2," and the products "go by odd and even."
>
> The sums are all rectangles, and the products are all squares.
>
> The difference between consecutive products increases by two.

The video also captures some students attempting to find patterns in the arrangement of numbers, which are most likely arithmetic relationships independent of the relationships between sums and products. For example:

> One student notices that 3 + 3 = 6 and that 3 + 6 = 9. It is likely here that the student is noticing how multiples of 3 work in the single number column.
>
> Another student notices that 2 + 2 = 4, 3 + 1 = 4, and 16 − 10 = 6. Although it is difficult to tell, it is likely that his "patterns" are arithmetic relationships that he has noticed based on numbers that happen to be near each other. (The student notices that when numbers are added, the answer is "down" [lower in the chart] and when numbers are subtracted, the answer is "up" [higher in the chart].)

After students have had time to work in small groups together, Ms. Doolittle convenes a whole-group discussion. She asks groups to share a pattern or a relationship that they found. The boy who was noticing that 2 + 2 = 4, 3 + 1 = 4, and 16 − 6 = 10 shares his observations. It seems apparent that Ms. Doolittle does not follow what he was describing and asks him to show his patterns on the chart. When he restates them using the chart, it becomes more evident that he is noticing arithmetic relationships of numbers that are near each other on the chart.

Ms. Doolittle does not pursue this student's observations; instead she moves on to another group. One student in that group explains that the sums grow one square (row of squares) longer going down to the next number, and the products grow one square (row of squares) longer going down and one square (column of squares) wider going to the next number. Ms. Doolittle probes for more information about sums, and the student adds that they are all even numbers.

Ms. Doolittle calls on another group, and Michael comes up and describes the growth pattern of the squares, noting that the difference between each product increases by 2. He points to these increases in the arrays:

> All right, you start with the 1 and then in the top left hand corner you put the 1 and then in between 1 and 4, there's 3 and then there's three [blocks] surrounding the one [block]. And then go down to the 9 and in the top left hand corner take out the four blocks and in between 4 and 9 there's 5 and there's five [blocks] surrounding the four blocks.

Ms. Doolittle prompts the students to look at their own charts as Michael repeats his observations of growth patterns. At this point in the lesson, Ms. Doolittle tells the students to write a journal entry about why they like looking for patterns and relationships.

A Note on Gender Balance As you or participants watch this video, you may notice that many of the students who share observations are boys and only a few are girls. Only boys are seen sharing in the whole group discussion at the end of the clip. If this issue comes up in the session, it might be helpful to note that this is an edited tape that captures only 15 minutes of a whole lesson. The evident gender imbalance may be the unintended consequence of editing choices as much as it might be a reflection of decisions made by the teacher. The limited data do not allow participants to know what considerations are guiding the on-the-spot pedagogical decisions Ms. Doolittle is making. After noting these things, mention to participants that later in the session they will have the opportunity to think about questions to ask Ms. Doolittle in the context of a post-observation conference and that they might want to include a question that helps them understand Ms. Doolittle's thinking about these issues.

Discussion of the First Viewing

Many educators respond positively to their initial viewing of this video. They see a class in which students are engaged in their explorations and eager to share their thinking. If this is the response of the participants in your seminar, you want to both honor what this teacher makes possible in her classroom and help participants draw on the mathematical exploration they just completed. Ask participants to consider what deeper understandings of growth patterns of sums and products (and of the power and role of prediction in mathematical reasoning) might have taken place, and what the teacher would need to have understood and done in order for these to emerge. Try to keep the discussion on students' ideas and the opportunities they might have provided for the teacher to help students make meaning of important mathematical ideas. Participants will already be familiar with some of these mathematical ideas from having just explored the activity. However, it may still be challenging for them to follow students' reasoning, particularly because the videotaped clip shows only small samples.

♦ What mathematical ideas are students grappling with in this investigation? Participants may give a range of responses to this question. You should pay particular attention to how mathematically specific they are able to be. Some participants may think that it is sufficient for students to develop skills in finding patterns. Although developing such skills is important, encourage participants to articulate why they are important. For example, a powerful idea behind finding patterns is *predictability*. Students who develop the skill of systematically investigating numerical patterns and relationships are developing the groundwork for being able to derive generalized mathematical statements and formulas. In this specific activity, for instance, understanding the pattern for how the square numbers increase is critical in being able to *predict* the square of any number when you know the square of one number.

With regard to the specific mathematics of this activity, although Ms. Doolittle stated at the beginning that she wanted her students "to try and investigate a relationship between multiplication and addition" (i.e., products and sums), participants might be puzzled about what the intent of the activity really was. Did Ms. Doolittle want the students to find a *particular* relationship? If she wanted students to find the growth patterns, why did she have them start by investigating single numbers? Whether she intended it that way or not, Ms. Doolittle's original question pointed students more towards looking for a relationship between addition and multiplication within a given row (i.e., for any given number added to or multiplied by itself), rather than attending to features of growth patterns within addition and multiplication that the representations exhibited.

In order to attend to the comparative growth patterns of addition (doubles) and multiplication (squares), it might have been interesting if the teacher had probed further into one student's comment that the shapes in the addition column are all rectangles and the shapes in the multiplication column are all squares. Why is that? Given that observation, which is growing faster? Do all multiplication problems, when built as arrays, come out as squares, or is there something particular about multiplying a number by itself that makes it a square?

Participants might also be curious about what Ms. Doolittle wanted to learn—or wanted the students to learn—by writing why they liked patterns. How would writing on that topic help them develop insights and deeper understandings into the relationship between addition and multiplication?

In this episode we see a teacher who has her students conduct an investigation of relationships and patterns in consecutive numbers, their doubles, and their squares. She gives them free rein to try out different patterns to see what they can discover. She supports them in expressing their ideas, and she listens to what they have to say. As the discussion of the intellectual community in this classroom should uncover, Ms. Doolittle is incorporating important components of standards-based practice into her classroom.

But the class discussion rarely reaches into important ideas about how addition and multiplication function. What would the teacher need to know and understand in order to bring these ideas out? How much value might have been added to this lesson had Ms. Doolittle had the opportunity, before teaching the lesson, to talk through the ideas about addition and multiplication that are important for fourth graders to build and to think about how this activity might help them do so? If she had been in the position to consider the algebraic concepts, if not notation, embedded in this activity beforehand, how might Ms. Doolittle have framed the problem differently for her students?

♦ How are students making sense of what it means to search for patterns and relationships? Even in this short excerpt, we see a range in students' grasps of searching for patterns. Some students are focused primarily on seeking arithmetic relationships and are using proximity as a guide. It may be possible that these students are relying on their previous experience with finding numerical patterns in number charts and are not yet comfortable with the idea of generalization. Others, however, are beginning to make a shift from looking for relationships between isolated numbers to looking for patterns that are consistent across numbers.

Discussion of the Second Viewing

For the second viewing, because of time constraints, the video is edited to show only a small segment of the lesson. When making the transition to discussing the intellectual community aspects of the class, be careful to keep the discussion linked to the mathematical content. Many participants can have a difficult time attending to mathematical content when they are asked to consider the intellectual community aspects of a classroom. They tend to switch their focus to interaction and process. They may consider such aspects as whether it is a safe setting for students or whether the teacher respects all students' contributions. Yet, an *intellectual* community is one that allows for the exploring of substantive ideas. It is important that you help participants see how an intellectual community helps students explore mathematical content.

The questions on the Intellectual Community Observation Guide are meant to help participants become more discerning about what they see in the mathematics classrooms and to support them in noticing elements that are or are not present. However, these questions are meant more to help participants raise questions than to evaluate or assess a classroom or a teacher.

In preparation for this part of the discussion, you should review the subsection "Orientation Toward Teachers When Observing" in the **Introduction,** which emphasizes the need to think about the teacher nonjudgmentally, focusing on what the teacher appears to understand and what she may still be working on. In addition, it will help participants continue to develop an orientation of inquiry toward classrooms if they raise their issues in the form of conjectures about what the teacher does or does not understand. For example, a participant might say:

> Ms. Doolittle asked the students to investigate a relationship between multiplication and addition (citing data). But this direction didn't seem to help the students focus clearly enough on the mathematics embedded in the columns and arrays. I have a conjecture that Ms. Doolittle had not actually worked out the mathematics for herself, and so was not aware that her question was too broad.

Such an orientation of inquiry toward the teacher is more concrete, more in line with the uncertainties of classroom observation, and more respectful of the possibilities for teacher learning than would be a statement such as, "Ms. Doolittle should have done the mathematics first." Additionally, a conjecture can be checked with the teacher in the context of a post-observation conference.

- **How are students showing respect for one another's ideas?** In small groups, there are several examples of students using one another as resources to build on their own insights into patterns. For example, at one point one boy says, "See, $3 \times 3 = 9$, $4 \times 4 = 16$, $5 \times 5 = 25$, $6 \times 6 = 36$, $7 \times 7 = 49$, $8 \times 8 = 64$, $9 \times 9 = 81$. What do you have, Justin?" and Justin comments that "3 plus 3 is 6 and then it skips 2." The boy makes a connection when he says, "And if it kept going on, it would keep skipping 2." Amber then says, "Yeah, $6 + 6$ and $6 + 3$ it would skip 2 (pointing to the numbers 7 and 8) and go to 9."

Although students were able to work well together, it seems that a number of students were using the table more as a number chart and looking for arithmetic relationships rather than looking at the properties of addition and multiplication, some of which the chart illuminated.

♦ **How does Ms. Doolittle support students in showing respect for one another's ideas?**
Ms. Doolittle has succeeded in creating an atmosphere of respect in her classroom. She calls on a variety of students to share, and she does not comment about the validity of students' responses after students share their thinking. For example, her expression and behavior toward the student who pointed out random number facts, such as $16 - 10 = 6$, was very similar to that toward the student who pointed out the growth pattern of successive squares and doubles. Although she asked the student who discovered a numerical relationship between successive squares ($1 + 3 = 4$, then $4 + 5 = 9$, and so on) to explain his thinking more fully, she gave no indication that his observation was more meritorious or interesting than the others. During the introduction of the activity, Ms. Doolittle waits patiently when students need time to fully articulate their ideas, and does not cut them off, rush them, or finish sentences for them. Students follow her example and rarely talk over one another, even when excited during group work.

Students are polite and respectful in their interactions with one another, and we see some examples of collaboration, such as the two boys noted above who work together. In the short segments available to us, it is hard to see a number of ideas being presented in depth. Still, we can see hints of this in the way Ms. Doolittle draws out ideas, as she did with the student who noticed growth by extra blocks in both the sum and product columns, and in the way she encourages students to try to understand others' ideas, as she did during Michael's presentation.

Ms. Doolittle takes an active role in addressing and drawing out ideas her students are developing. Still, she might have pushed her students' thinking further by encouraging them to comment on, respond to, or build on one another's ideas as they were being presented to the class. It might also have been productive for her to ask students to come up with reasons why their observed patterns might work and to consider whether these patterns would have worked for any number. Teachers are always juggling alternative teaching moves, and because we do not know what was in Ms. Doolittle's mind, we cannot be aware of what she might have been weighing as she made pedagogical decisions. It is important for participants to recognize that there will always be potential paths not followed.

♦ **How do students use one another as resources as they make sense of mathematical ideas?** As was discussed previously, some students build on one another's ideas in noticing the growth pattern of sums. Justin comments that "3 + 3 is 6 and then it skips 2." Another student makes a connection and says "Yeah, it keeps on skipping," and another says, "And if it kept going on, it would keep skipping 2." In this same conversation, one student seems to misunderstand what the others were observing, as Amber then says "Yeah, 6 + 6 and 6 + 3 it would skip 2 (pointing to the numbers 7 and 8) and go to 9." The other students commenting on the growth pattern they noticed sparked her observation, that there are two digits between 6 and 9. However, this observation was not mathematically related to the growth pattern of sums. Much of the time students are sharing ideas but not really building on one another's ideas.

Some participants may comment on this and note that there were seemingly unproductive comments made alongside productive mathematical exploration. In spite of these apparent side trips, the students do generate valid mathematical observations concerning the growth patterns of both sums and products. Even the group that chose to present random arithmetic relationships had also generated a number of observations about growth patterns, including that "numbers go by one, sums go by twos, and products go by odd, even."

In another example of students building on one another's thinking, we notice a boy sharing an early observation of the numerical relationship between successive squares during small group work. He says:

> Zero plus one is one, and there's one difference between zero (imagining a previous, not visible step in the pattern) and one (the first square number). One plus two is three; there's three difference between one and four (the next square number). Okay? Then there's three plus two is five, and there's five difference between four and nine (the next square number). Then five plus two is seven, and there's seven difference between nine and sixteen.

Later, when we see this group represented by Michael present its results at the board, the group has refined its observations to notice that the differences, the successive odd whole numbers, can be arranged around each preceding square to make the succeeding square. Michael notes this as follows:

> All right, you start with the 1 and then in the top left-hand corner you put the 1 and then in between 1 and 4, there's 3 and then there's three [blocks] surrounding the one [block]. And then go down to the 9 and in the top left hand corner take out the four blocks and in between 4 and 9 there's 5 and there's five [blocks] surrounding the four blocks.

This outcome seems to be an extension of the group working together to build and develop their ideas. The fact that a different student initially noticed the relationship than the student who presented the work to the class demonstrates the exchange of knowledge that took place in this group.

♦ **What questions about student thinking that took place in this episode might be fruitful to discuss with the teacher?** This question is meant to prefigure the "co-inquiry" strand of the course, which will be discussed in Session 5. It is intended to start participants thinking about how they could have a conversation with a teacher in which both they and the teacher are learners. For many participants, this will be a challenging shift because they are used to being in positions of authority vis-à-vis teachers. Your goal is to help participants generate questions related to student thinking that they are genuinely curious about rather than raising questions they think they *should* ask because it is their responsibility to do so.

You will need to be particularly sensitive to the ways in which participants interpret and respond to the question of what they might like to discuss with the teacher. Many administrators have a tendency to shift their mindset from being reflective or investigative to being critical of the teacher: "I would want to know why she didn't ask the students this" or "I would want to know why he didn't use this type of question." This section, then, is intended to help participants think about what they might be curious about, what they didn't understand or are puzzled about, or what they might like to find out. This is the beginning of participants' exploration of collaborative inquiry as a way of encouraging teachers to develop curiosity about the mathematical thinking of the students in their classes.

Examples of co-inquiry questions might include the following:

> I'm wondering what different students actually thought about what a mathematical pattern is. Michael seemed to be on to a significant pattern, but the boy before him seemed to be pointing to isolated number facts.
>
> I'm wondering how collaborating seemed to help some students identify mathematical patterns but perhaps limited others. I'm thinking about the boy who was dismayed because he collaborated but didn't get the right answer.

Remember that co-inquiry questions are not intended to be questions that challenge the teacher's competency or make the teacher defensive about what happened. Rather, they are intended to spark discussion about interesting features and outcomes of the lesson.

When participants discuss what they would want to talk about with this teacher, you will need to help them pay attention to the substance of what they say. Administrators often focus on *how* to talk with teachers: how they should phrase their questions, how not to sound threatening, and so on. These are critical issues in cultivating trust with teachers, but it is also crucial to understand *what* can be productive content in a conversation with a teacher.

CLOSING
20 minutes

Materials
Handout 14

Homework

10 minutes

Assigning Homework As you assign the homework, you might mention that this was the second of two sessions on intellectual community in the classroom. The next two sessions will focus on observing the mathematical content of a mathematics class. The first reading assigned for homework, *Approaches to Solving a Fair-sharing Problem* (Reading 4), will provide administrators with the opportunity to learn about some of the important ideas about fractions in the elementary curriculum.

Let participants know that prior to reading this first article, they will work on the same math problem that is the focus of the reading. By doing the math themselves beforehand, they will be better able to understand the mathematical thinking of the students whose approaches are featured in the reading.

For the second reading, *Magical Hopes: Manipulatives and the Reform of Math Education* (Reading 6), participants think about how the relationship between mathematical ideas and the way they are represented affect students' learning.

Bridging to Practice

10 minutes

Reflective Writing Remind participants that Bridging to Practice is designed to allow them to focus on the ideas in the course that seem particularly interesting for them and for their school community. Point out that it provides a framework for planning what they can do in order to move themselves and their schools along with these ideas.

Post the following questions, and invite participants to respond to them:

- *Choose an idea that came up today that you found particularly interesting. What is your current thinking about this idea?*
- *Where is your school now with regard to this idea?*
- *What are one or two things that you, as instructional leader, will pursue to move yourself and/or your school along with this idea?*

Your Own Journal Writing

Within one or two days of teaching this class, after you have had the chance to unwind from the class and perhaps talk with a colleague or friend about how it went, set aside some time to write your own journal entry about the class. We encourage you to write about the ideas held by the participants in your group rather than your own actions and sense of how things worked out. In our experience, writing about what the participants are thinking about will be much more useful to you as you prepare for the next session. Some facilitators use the time when administrators are writing in their own journals to make preliminary notes for this journal writing.

Products and Sums Pattern Exploration

This activity explores the relationships between products and sums. Below is a table with the numbers 1–10 and two blank columns, one for the double of the number and other for the square of the number.

Number	Sum (2x)	Product (x^2)
1		
2		
3		
4		
5		
6		
7		
8		
9		
10		

1. Take a few minutes to fill in the chart on your own. As you are filling it in, write down any patterns or relationships that you notice or would like to explore further.

2. Observing Growth Patterns:
   a. After you have filled in the chart, you will be placed into groups of two or three. Using graph paper or tiles, create a visual representation of the numbers in each column. Look carefully at the numbers and the representations of them. What do you notice? How are the columns similar, and how are they different?

b. Look specifically at the products column and your representation of it. How can you describe the growth pattern of the numbers? How does it differ from the growth pattern in the sum column? What does this tell you about how multiplication works? How is it different from addition? Write a statement that describes the growth pattern of the products.

c. From your visual representation of products, write a statement to predict the next square number for any number (e.g., if you know the square of 7, how would you know the square of 8?). See whether you can find a way of describing the growth pattern in words and with an algebraic expression.

3. Observing the relationship between the each sum and product:
   a. Look at each row of numbers. What relationships can you find among any given number, its double, and its square?

   b. Use a diagram or numerical statement to illustrate any relationships that you find.

Observation Guide

Students		
Focus Question	Conjectures	Evidence from Classroom
Math Content		
• What mathematical ideas are embedded in the lesson?		
• What makes this worthwhile mathematics?		
Learning		
• What kinds of mathematics sense-making are students doing?		
• What mathematical ideas seem to be confusing to students?		
• In what ways can you see that the students are developing their mathematical ideas over time?		
Intellectual Community		
• How are students showing respect for one another's ideas?		
• How do students use each other as resources as they make sense of mathematical ideas?		
• What evidence beyond raised hands do you have that students are engaged?		

Observation Guide

Teachers		
Evidence from Classroom	Conjectures	Focus Question
		Knowledge of Content
		• What does the teacher seem to understand about the mathematics?
		• What is the teacher's long-term mathematical agenda?
		• What does the teacher seem to understand about the development of children's ideas in this topic?
		Pedagogy
		• How does the teacher work with the sense the children are making?
		• How does the teacher work productively with students' confusion?
		• How does the teacher attend to all students?
		• How does the teacher adjust her teaching based on the ideas she hears from students?
		Facilitating Intellectual Community
		• How does the teacher support students in showing respect for one another's ideas?
		• How does the teacher set the tone so students see each other as resources for mathematical thinking?
		• What interventions does the teacher make to ensure that students' engagement has a focus on mathematical ideas?

Intellectual Community Observation Guide

Students		
Focus Question	Conjectures	Evidence from Classroom
How are students showing respect for one another's ideas?		
• What are students doing to show that they are listening to other students?		
• How willing are students to share their ideas even if they know that they aren't correct?		
• How attentive are students to one another's ideas?		
How do students use each other as resources as they make sense of mathematical ideas?		
• Are they building on each other's mathematical ideas?		
• Are they asking each other questions related to mathematical ideas?		
What evidence beyond raised hands do you have that students are engaged?		

Intellectual Community Observation Guide

Teachers		
Evidence from Classroom	Conjectures	Focus Question
		How does the teacher support students in showing respect for one another's ideas?
		• How does the teacher set and maintain norms of interaction for discourse?
		• How does the teacher support such practices as: • attentive listening • question-posing about mathematical ideas • provisional thinking
		How does the teacher set the tone for students to see each other as resources for mathematical thinking?
		• Does the teacher invite students' tentative thinking?
		• Does the teacher invite students to build on one another's ideas?
		What interventions does the teacher make to ensure that students' engagement has a focus on mathematical ideas?

Homework for Session 4

1. The first assignment for Session 4 is to read an excerpt from *Young Mathematicians At Work: Constructing Fractions, Decimals, and Percents* by Catherine Twomey Fosnot and Maarten Dolk (Reading 4).

   This excerpt describes the approaches two pairs of students in a fourth/fifth grade classroom used to solve a fair-sharing problem.

   As you read the excerpt, work through the ways the students approached the problem. Also note what mathematical ideas seemed to be difficult or puzzling for students.

2. In the next session we will consider the Mathematics Content Observation Guide. In order to be ready to view videotaped clips from this perspective, it is important to distinguish between the mathematical *ideas* that students in a class are working with and the ways in which those ideas might be represented in instruction—with manipulatives, diagrams, or numbers.

   As a way of thinking about the relationships between mathematical ideas and their representations, please read *Magical Hopes: Manipulatives and the Reform of Math Education* by Deborah Ball (Reading 6). Identify three ideas in this article that are especially interesting to you, and write at least a paragraph about each.

SESSION 4

Observing for Content: Listening to Children's Ideas About Fractions

In this session we continue to explore how administrators can develop a new "eye" for mathematics classrooms, with a focus now on the mathematics content of the lesson. Administrators may be accustomed to paying only cursory attention to the mathematics content of the lessons they observe. They may take note of the topic of instruction and look to see whether the instruction includes use of manipulatives, small-group work, and so on. In this session, participants consider the importance of looking beyond the broad topic (e.g., fractions) to the mathematical ideas within that topic (e.g., equal shares do not need to be congruent) in order to grasp more fully the mathematical content of the class.

In standards-based classrooms, the mathematics under investigation and the teacher's knowledge of it are centrally important. Those who observe in classrooms need to appreciate that students' mathematical understandings develop slowly through the exploration of mathematical ideas in a variety of contexts. The pedagogical process—the process of helping students develop their subject-matter knowledge—is inextricably interwoven with that knowledge because teachers are working with students' mathematical ideas. In their observations, administrators will need to attend to the subject matter thinking that students are engaged in and to ways that teachers help students develop their thinking in order to make valid judgments about the quality of the teaching they observe.

These ideas will be the focus of investigation and discussion in the next two sessions. The Math Content Observation Guide will provide scaffolding for participants to attend to the math content and to the teacher's knowledge of the math content in the videotaped classroom episode.

The activities of this session are designed to bring out two important ideas: that children's mathematical ideas develop slowly over time, and that observers need to attend to children's mathematical knowledge and the teacher's approach to help them push their thinking farther.

Overview for Session 4

OPENING page 127 5 minutes	**INTRODUCTION** This is a time for announcements and session overview. It is also an opportunity to provide an overview of this session and its connections to the previous and upcoming sessions.
ACTIVITY 1 Homework Discussion page 129 35 minutes	**CONSIDERING THE COMPLEXITY OF FRACTIONS** Participants work in small groups to respond to questions about the mathematical thinking of two pairs of students from the Fosnot and Dolk reading. In answering these questions, participants gain an appreciation for some of the key ideas related to fractions and potential areas of difficulty for children in understanding these concepts.
ACTIVITY 2 Mathematics Exploration page 138 35 minutes	**THE FRACTIONS TRACKS GAME** Participants play the game "Fractions Tracks" from *Investigations in Number, Data and Space* that the students play in the classroom episode to be viewed in this session. This activity gives participants the opportunity to explore equivalence, an important aspect of fractions.
ACTIVITY 3 Video page 140 70 minutes	**VIEWING THE FRACTIONS TRACKS VIDEO** Participants watch the *Fractions Tracks* video and use the Math Content Observation Guide. As participants view the classroom episode, they will adopt a listening attitude in order to understand what ideas various students are working on. They will also gain an appreciation for the range of understanding of equivalent fractions among a group of students.
ACTIVITY 4 Homework Discussion page 149 15 minutes	**DISCUSSION OF MAGICAL HOPES** Participants discuss the article *Magical Hopes: Manipulatives and the Reform of Math Education* by Deborah Ball (Reading 6). Participants will be encouraged to push their thinking about the role of manipulatives in mathematics classrooms.
CLOSING page 151 20 minutes	**HOMEWORK** There are three homework assignments for the next session: read *Capturing Teachers' Generative Change: A Follow-up Study of Professional Development in Mathematics* by Megan Loef Franke, Thomas P. Carpenter, Linda Levi, and Elizabeth Fennema (Reading 7); use the Math Content Observation Guide in an actual classroom observation; and create a pie chart that represents the communication styles they employed in a recent post-observation conference. **BRIDGING TO PRACTICE** Participants finish the session with journal writing that allows them to develop further some of their thinking from the session.

Big Ideas

Participants explore:

- the complexity of fractions
- the development of students' understandings of fractions over time
- the importance of attending to mathematical content when observing mathematics classrooms

Materials

- ☐ Handout 14, Homework for Session 4
- ☐ Reading 4, "Approaches to Solving a Fair-sharing Problem" by Fosnot and Dolk
- ☐ Reading 6, *Magical Hopes: Manipulatives and the Reform of Math Education,* by Deborah Ball
- ☐ Handout 15, Considering the Complexity of Fractions
- ☐ Handout 16, Directions for Playing Fraction Tracks
- ☐ Handout 17, The Fraction Tracks Board
- ☐ Handout 18, The Fraction Tracks Cards
- ☐ Handout 19, Fraction Tracks Reflection Questions
- ☐ Chips, pennies, or markers of some sort
- ☐ Videotaped Clip 4, *Fraction Tracks* (DVD #2; Program 1, 0:54–14:57)
- ☐ Reading 5, *Fraction Tracks Transcript*
- ☐ Handout 20, Overall Observation Guide
- ☐ Handout 21, Math Content Observation Guide
- ☐ Reading 7, *Capturing Teachers' Generative Change: A Follow-up Study of Professional Development in Mathematics* by Franke *et al.*
- ☐ Handouts 22, 23, and 24, Homework for Session 5

Preparation

- Prepare an agenda for the session on an overhead or on flip chart paper with approximate times for each activity noted.
- Write the prompts for Bridging to Practice on an overhead or on flip chart paper.
- As you look over the discussion questions for the activities in this session, think about which ones your group would most benefit from considering, and use these to guide your discussions.
- Read the two homework articles for this session, Reading 4,"Approaches to Solving a Fair-sharing Problem," an excerpt from the Fosnot and Dolk book for Activity 1, and Reading 6, *Magical Hopes: Manipulative and the Reform of Math Education,* by Deborah Ball for Activity 4. Familiarize yourself with the discussion questions.

PREPARATION

- Play the Fraction Tracks game from *Investigations into Number, Data, and Space* played in Activity 2. (It would be fun for you to do it with a family member, a colleague, a friend, or a child. After playing the game, the two of you should have a conversation based on the discussion questions that go along with the Fraction Tracks game.) Make a copy of the Fraction Tracks board for each set of four participants and of the fraction cards (cutting apart copies of the enclosed sheet) for each pair of participants.
- Watch Videotaped Clip 4, *Fraction Tracks,* shown in Activity 3. View the tape several times until you can see the characteristics of student thinking that are described in the **Facilitator Notes** for Activity 3. Use the Math Content Observation Guide on a couple of the viewings of the videotape. Review the questions in the Observation Guide.
- Participants view the *Fraction Tracks* video twice, once using the Overall Observation Guide and the second time using the Math Content Observation Guide. As we advocate throughout this course, plan to leave enough time in the session for participants to view the videotaped clips two times. You may want to refer to the section *Viewing Videotaped Clips of Teachers and Students at Work* (p. 9) in the **Introduction** to remind yourself of the rationale that underlies this recommendation.
- Participants will be making a pie chart using Handout 23 for homework. Standardization of both the pie chart itself and the color coding for the different communication styles is important for this activity to make it possible for participants to make connections among the various charts in the next session's discussion.
- Review participants' writings, paying particular attention to how they are making sense of the notion of the intellectual community in the mathematics classroom, in particular the role of the mathematics itself in such a community. This will help you decide what topics to emphasize in the discussions for this session.

OPENING
5 minutes

> ***Materials***
> Name tags
> Agendas

Introduction

Before starting the session, post the agenda in a place where everyone can see it.

5 minutes

Introducing the Session Let participants know that, as part of developing a new "eye" for mathematics classrooms, they will attend to the mathematics content of the lesson for the next two sessions. Later in this session they will work closely with the Math Content Observation Guide.

Overview

By the time you meet with your group for the fourth session, you will probably feel that the group has started to "gel." Participants will know each other, and there will be a history of discussions from previous sessions that you can draw on and refer back to.

In this session, the focus shifts from the intellectual community of the classroom to the mathematical content of the lesson. Participants consider the questions in the Math Content Observation Guide that they will be thinking about when observing in a mathematics lesson. These questions attempt to ascertain the sense students are making of the mathematics being presented and the way the teacher is working with these mathematical ideas. In the **Facilitator Notes,** the central mathematical ideas of the articles participants read and videotaped clips shown are highlighted to help you gain some depth of understanding of the mathematics so that you can follow the ideas brought up by participants or introduce ideas that the participants might not be thinking about. Remember that the discussion questions are intended primarily to start discussion. As facilitator, you want to listen for the ideas that participants present in response to the questions and direct the discussion so that it is most productive for the group.

ACTIVITY 1
Homework Discussion
35 minutes

Materials
Handout 15

Considering the Complexity of Fractions

20 minutes
Small group

Discussing the Fosnot and Dolk Reading Let participants know that the excerpt they read for homework is from a series of books by Catherine Twomey Fosnot and Maarten Dolk about how children construct their knowledge of mathematical concepts. The book from which the reading is drawn focuses on the thinking of children in grades 5–8 about fractions, decimals, and percents. The other two books, *Young Mathematicians at Work, Constructing Number Sense, Addition and Subtraction* and *Young Mathematicians at Work, Constructing Multiplication and Division* are about the mathematical thinking of children in preK–3 and grades 3–5 respectively.

Tell participants that they will be taking a close look at the mathematical thinking of the two pairs of students to determine how they are making sense of the mathematics. Pass out Considering the Complexity of Fractions (Handout 15), and have participants pair up to discuss the questions. If you have paired participants ahead of time, indicate with whom they will be working.

By making sense of students' thinking about the mathematics, participants can gain insight into both the mathematics *and* children's thinking about the mathematics.

There are four questions on the handout, and because it is likely that there will not be enough time for participants to reflect on all of them in depth, suggest to groups that they each begin with a different question. This will ensure that during the whole group discussion, at least one group will have reflected in depth on each of the questions.

The questions on the handout are as follows:

1. *When Ms. Mosseson asks her students whether all groups had the same amount to eat, some of her students respond that the students in the three groups where there was one fewer submarine sandwich than students would each have the same amount to eat [lines 29–47]. What seems to be going on for the students who respond this way?*
2. *What are Ernie and Jackie puzzling about when they are trying to determine how much the Museum of Modern Art group got to eat? [lines 106–132]*
3. *Compare the strategies that Jackie and Ernie on the one hand [lines 106–132] and Nicole and Michelle on the other [lines 74–86] use to figure out how much the students on the Museum of Modern Art field trip got to eat. What ideas about fractions emerge from a comparison of these approaches?*
4. *What do Jackie and Ernie need to understand in order to know that the Museum of Modern Art group had less to eat than the Ellis Island group? [lines 165–172]*

As pairs are working, move around from group to group to gather information about the thinking of your groups to inform your decisions about the whole group discussion that follows.

15 minutes

Whole group

Discussing the Reading Rather than focusing on each of the questions on the handout, the whole-group discussion questions ask participants to look more globally at the learning and teaching of fractions in the context of this class as students work on a fair-sharing problem. If particular issues came up as the groups were working together on the handout, you might want to take a few minutes at the beginning of this discussion to address them.

The questions for the whole group discussion are as follows:

> *What ideas about fractions surfaced as students worked on this problem?*
>
> *What is complex about fractions?*

As participants discuss this, ask them for specific examples from the homework reading to support their ideas.

> *What was Ms. Mosseson's approach to working with her students?*

Because participants get only limited information of how Ms. Mosseson works with her students' ideas, let them know that following the small group work on the fair-sharing problem, she holds a whole-class discussion in which she chooses groups to present their work. She selects groups whose approaches and strategies are likely to illuminate an understanding of the important mathematical ideas of the lesson and from whose thinking other students would benefit.

Let participants know that in the next activity, they will focus in some depth on one of the central ideas about fractions that came up in Ms. Mosseson's class—that of equivalence.

The first question helps participants see that there are several different concepts for students to understand about fractions.

The second question allows participants to consider what concepts related to fractions can be challenging for students.

The third question focuses participants' attention on the role that the teacher assumes in this initial part of the investigation.

Overview

Fractions are a complex topic and one that takes time for children to come to understand. Children quite naturally develop a number of ideas about fractions, and they need many opportunities to explore fractions, use fractional language, and learn to represent fractions with standard symbols. Although it is not an expectation that participants in your group will come away with an in-depth understanding of fractions, participants should realize the complexity of fractions and appreciate what concepts can be hard for children to understand about fractions.

Because this is the first of two sessions in which participants concentrate on the Mathematics Content Observation Guide, they need to spend a good deal of time making sense of the mathematical ideas themselves. In the first activity, participants deepen their understanding of the mathematics in the lesson and of students' mathematical thinking by talking through the approaches two pairs of students take to solving a fair-sharing problem.

Summary of the Fosnot and Dolk Reading

The excerpt participants read for homework is from *Young Mathematicians at Work, Constructing Fractions, Decimals, and Percents,* one of three books by Catherine Twomey Fosnot and Maarten Dolk that focus on how children construct their knowledge of mathematical concepts. Participants may be interested in knowing about the other two books, *Young Mathematicians at Work, Constructing Number Sense, Addition, and Subtraction,* and *Young Mathematicians at Work, Constructing Multiplication and Division.*

The reading begins with a description of a fair-sharing problem posed by Carol Mosseson, the teacher in a fourth/fifth grade classroom in New Rochelle, NY. She asks her students to determine whether all of the students in her class last year who took field trips to museums around New York City received equal shares of the submarine sandwiches prepared by the school's cafeteria. The students attending the museums were in groups of unequal sizes, and each group had a different number of submarine sandwiches to share among its members.

> The four students who went to the Museum of Natural History had three subs to share.
>
> The five students who went to the Museum of Modern Art had four subs to share.
>
> The eight students who went to Ellis Island and the Statue of Liberty had seven subs to share.
>
> The five students who went to the Planetarium had three subs to share.

Museum	# Students	# Subs
Natural History	4	3
Modern Art	5	4
Ellis Island	8	7
Planetarium	5	3

When Ms. Mosseson poses this problem to her class, the initial reaction of some students is that the students in the three groups in which there was one fewer sub than students would each have the same amount to eat. Students also note that the group with five students and three subs would have less to eat than the students in the other groups, but if this group had gotten one more sub, everyone would have gotten the same amount to eat.

Other students in the class support an alternative way of thinking about this—that the sizes of the pieces across groups would not be the same, and therefore students on the field trips would not have gotten the same amount to eat.

Ms. Mosseson asks her class to form groups to investigate the problem further, guided by two questions:

How much did each child in each group get, assuming that the subs were all shared equally in each group?

Which group got the most?

Jackie and Ernie's Thinking, Part 1

♦ **Natural History Museum (4 students, 3 subs)** Jackie and Ernie draw a picture of how the subs would be divided for each group (p. 51 in the book of *Readings*). They reason that because there were three subs for four students, they would cut two subs in half and the third sub into fourths, giving everyone a half sub plus one fourth.

♦ **Ellis Island (8 students, 7 subs)** They use the same approach as they used for the Natural History Museum. That is, as Ernie explains, they gave each student a half sub, which uses up four of the seven subs. They then cut one of the three remaining subs into eight pieces, and they cut the other two subs into four pieces each. Thus, each student got one half plus one eighth plus one fourth. (Note that their drawing shown in Figure 1.2, p. 51 in the book of *Readings,* shows something different from what Ernie articulates. In the drawing, each of the three remaining subs is split into eight pieces.)

Ms. Mosseson notes that this pair's strategy is to make *unit fractions;* that is, fractions with numerators of one.

Museum	# Students	# Subs	Amt. Each Student Got
Natural History	4	3	$\frac{1}{2} + \frac{1}{4}$
Ellis Island	8	7	$\frac{1}{2} + \frac{1}{8} + \frac{1}{4}$

Nicole and Michelle's Thinking

♦ **Natural History Museum (4 students, 3 subs)** Nicole and Michelle also use a unit fraction strategy to figure out how much the each of the four students with three subs had to eat.

♦ **Museum of Modern Art (5 students, 4 subs)** For the Museum of Modern Art situation, Nicole and Michelle change their strategy and divide each of the four subs into five pieces because there were five students, giving each student a fifth of each sub. Nicole observes, "That's four fifths of a sub for each kid because it is four times one fifth."

♦ **Ellis Island (8 students, 7 subs)** Nicole and Michelle go back to using unit fractions for this part of the problem. Ms. Mosseson thinks about suggesting that they use the same approach they used on the Museum of Modern Art situation. She decides against suggesting this and speculates that there may be something about the context of each of the four scenarios that dictates the approach her students take. She wonders whether her students think that although dividing a sub into halves, fourths, and fifths is a reasonable thing to do, cutting subs into eighths should be avoided unless absolutely necessary. That would explain why for the Ellis Island group, Nicole and Michelle decide *not* to divide each of the eight sandwiches into eight parts; rather, they switch back to the use of unit fractions and give each student one half, plus one fourth, and only then, because there is just one sub sandwich left to be divided among the eight students, do they resort to dividing it into eights.

Museum	# Students	# Subs	Amt. Each Student Got
Natural History	4	3	$\frac{1}{2} + \frac{1}{4}$
Modern Art	5	4	$\frac{4}{5}$
Ellis Island	8	7	$\frac{1}{2} + \frac{1}{4} + \frac{1}{8}$

Jackie and Ernie's Thinking, Part 2

♦ **Museum of Modern Art (5 students, 4 subs)** Jackie and Ernie first cut three of the subs in half and give everyone a half, leaving one and a half subs still to be divided among the five students. They then take the whole remaining sub, divide it into five parts, and give each student one fifth. Lastly, they cut up the remaining half into five parts. When Jackie and Ernie describe how much each student in this group got to eat, they reason that it would be one half, plus one fifth, and then they puzzle about what to call the final piece which is a fifth of the half of a sub. Initially they refer to it as a fifth, but then they notice that these fifths appear to be half the size of the fifths of a whole sub. Ernie refers to it as a "half of fifth" and wonders what that is called. Jackie talks about it as a "fifth of a half," and then they conclude, "If it's a half of a fifth, then it must be a tenth." Ernie builds on this idea by suggesting that if they cut the other half into fifths also, they would have tenths.

♦ **Planetarium (5 students, 3 subs)** Here they cut the subs into halves and give each student one half, leaving one remaining half to be divided into five pieces. They then conclude that each student gets one half plus one tenth.

Museum	# Students	# Subs	Amt. Each Student Got
Modern Art	5	4	$\frac{1}{2} + \frac{1}{5} + \frac{1}{10}$
Planetarium	5	3	$\frac{1}{2} + \frac{1}{10}$

Jackie and Ernie's Thinking about Which Group Got the Most to Eat

The first thing Jackie and Ernie do to compare how much each group got is to eliminate the halves from each group because everyone in each group got a half. They then set out to compare the remainders.

♦ **Comparing remainders for the Ellis Island and Modern Art groups** When they examine the remainders, they determine that each member of the Ellis Island group had one eighth of a sub more to eat than the Natural History group, and each member of the Modern Art group had one fifth of a sub more to eat than the Planetarium group.

Museum	Remainders
Natural History	$\frac{1}{4}$
Modern Art	$\frac{1}{5} + \frac{1}{10}$
Ellis Island	$\frac{1}{8} + \frac{1}{4}$
Planetarium	$\frac{1}{10}$

♦ **Comparing remainders for the Natural History and Modern Art groups** Jackie and Ernie first decide to find a number that is divisible by the denominators (5, 4, and 10), and they choose the number 20. Next, they put 20 unifix cubes together and determine that a one-fourth piece is equal to five cubes. They then turn their attention to the Museum of Modern Art group, and figuring that one fifth is the same as two tenths, they add another tenth to this to get three tenths. To get the number of equivalent cubes, they reason that one tenth of 20 cubes is two cubes, so three tenths equals six cubes.

Museum	Remainders	Amt. of sub in unifix cubes
Natural History	$\frac{1}{4}$	5 cubes
Modern Art	$\frac{1}{5} + \frac{1}{10}$ or $\frac{3}{10}$	6 cubes

♦ **Comparing remainders for the Ellis Island and Modern Art groups** To do this, Jackie and Ernie look at the Ellis Island data and reason that a one-fourth piece of sub is equal to two one-eighth pieces, so each member of that group had three eighths of a sub ($\frac{2}{8} + \frac{1}{8}$) They conclude that because eighths are bigger than tenths, the amount of sub that the Ellis Island group had ($\frac{3}{8}$) was larger than the amount of sub the members of the Modern Art had ($\frac{1}{10} + \frac{1}{5}$, or $\frac{1}{10} + \frac{2}{10} = \frac{3}{10}$). For the Planetarium group, Ernie and Jackie reason that they had the least to eat because tenths are smaller than fourths or fifths, and the more pieces you have, the smaller each piece is.

The Mathematics in the Submarine Sandwich Problem

There are a number of key ideas related to fractions that students in Ms. Mosseson's class consider.

The Whole Matters When Comparing Fractions

In asking her students to solve the problem as presented, Ms. Mosseson helps them construct an understanding of the idea that in order to compare fractions, the wholes must be the same. Some initial responses are that groups whose numbers exceed the number of subs by one all got the same amount to eat (4 students with 3 subs, 5 students with 4 subs, and 8 students with 7 subs). The students who propose this do not seem to be aware that because the number of students in the groups differs, the pieces of the subs each member got to eat would differ in size.

Part/Whole Relations

The idea that fractions take their value from the whole of which they are a part is also central to an understanding of fractions. Jackie and Ernie are grappling with this idea when they encounter the idea that the fraction with which they are working, fifths, can relate to different wholes. At one point, they use $\frac{1}{5}$ to refer to a fifth of a whole sub, but at another, they use it to refer to a fifth of a half of a sub, which they eventually decide is the same as one tenth.

Equivalence

Another significant idea that comes up is that of equivalence. Ernie and Jackie make use of the idea of equivalence when they compare remainders for the Ellis Island and Modern Art groups.

First they figure that one fifth is the same as two tenths, and they add another tenth to this to get three tenths. Then they determine that a one-fourth piece of sub is equal to two one-eighth pieces, and thus each member of that group had three eighths of a sub. They are now in a position to compare the amounts that the Modern Art and Ellis Island groups got to eat by reasoning that $\frac{3}{10}$ would yield smaller pieces than $\frac{3}{8}$, so the Modern Art group got less to eat. Jackie and Ernie are also working with the idea of the smaller the denominator, the bigger the piece.

Museum	Remainders
Natural History	$\frac{1}{4}$
Modern Art	$\frac{1}{5} + \frac{1}{10}$ or $\frac{3}{10}$
Ellis Island	$\frac{1}{8} + \frac{1}{4}$ or $\frac{3}{8}$
Planetarium	$\frac{1}{10}$

An additional key mathematical idea, one that is not germane to fractions only, emerges during the interaction between Ernie and Jackie about what to call one fifth of a half of a sub. Here Ernie observes that one half of a fifth is the same thing as one fifth of a half. Although Ernie may not be thinking about multiplication here, it is the case that his observation is related $\frac{1}{2} \times \frac{1}{5} = \frac{1}{5} \times \frac{1}{2}$, or the commutative property of multiplication.

Developing an understanding of the commutative property of multiplication is not central to this lesson.

Equivalency vs. Congruency

Another idea that is central to this fair-sharing problem is that fractional pieces do not have to be congruent to be equivalent. (Participants encountered this same idea in Session 1 in the context of the classroom episode, *Fractions with Geoboards.*) As the students in Ms. Mosseson's class cut up the subs and move the pieces around to figure out how to share them fairly, they are exploring the idea that although the pieces have different sizes, the quantity stays the same. This concept is embedded in the two different strategies for fair-sharing that Jackie and Ernie on the one hand, and Nicole and Michelle on the other, use to solve the problem. Ernie and Jackie use a *unit fraction* approach to get one half plus one fifth plus one tenth, and Nicole and Michelle divide each of the four subs into five pieces to get the equivalent, but not congruent $\frac{4}{5}$ of a sub each.

Connecting Multiplication and Division to Fractions

In solving the problem, students in Ms. Mosseson's class also consider how fractions are connected to both multiplication and division. This idea comes up in the context of Nicole and Michelle's approach to figuring out how much five students sharing four subs would get. They explain that they divided each sub into fifths because there were five students. Nicole, pointing to each fifth, says that each student gets four fifths of a sub because it is four times one fifth. In other words, they are introducing the idea that 4 divided by 5 is the same as 4 times $\frac{1}{5}$.

Finding a Least Common Multiple When Comparing Fractions with Different Denominators

When Jackie and Ernie are comparing remainders for the Natural History ($\frac{1}{4}$ of a sub) and Modern Art ($\frac{1}{5} + \frac{1}{10}$ of a sub) groups, they look at the denominators to come up with a number that is divisible by five, four, and ten. They decide to use 20 as their common denominator.

Small Group Discussions

Participants work with a partner to think through the discussion questions on Handout 14. You may want to group those participants who have had more experience in mathematics with partners who have not for this discussion.

As a way of helping participants understand the mathematical thinking of the students featured in the excerpt, have them work through the mathematics themselves. Their own work with the mathematics will not be discussed separately during this session, but it will likely inform the discussions participants have about the students' approaches.

The questions on Handout 14 are meant to encourage participants to deepen their own understanding of the fractions in the context of a fair-sharing problem and to develop an understanding of the mathematical thinking of students.

1. **When Ms. Mosseson asks her students whether all groups have the same amount to eat, some of her students respond that the students in the three groups where there was one fewer submarine sandwich than students would each have the same amount to eat [lines 29–47]. What seems to be going on for the students who respond this way?** This question asks participants to tap into the thinking of the students who seem to propose solutions based on the difference between the number

of subs and students in each of these groups, which is one. These students are not taking into consideration that each group has a different number of members. Because the totals or wholes are different, the sizes of the pieces each group gets would be different.

2. **What are Ernie and Jackie puzzling about when they are trying to determine how much the Museum of Modern Art group got to eat? [lines 106–132]** Ernie and Jackie are trying to work out what the whole is in relation to the parts. When they divide a whole sub into five parts, they treat the whole as one whole sub and each part as one fifth. Then, when they split a half sub into five parts, initially they do not take into account that this time they are dealing with a half sub, and they refer to these pieces as fifths as well. When they realize that one fifth of a half sub is much smaller than one fifth of a whole sub, they come to see that the pieces are both fifths of a half sub and tenths of a whole sub.

3. **Compare the strategies that Jackie and Ernie on the one hand [lines 106–132] and Nicole and Michelle on the other [lines 74–86] use to figure out how much the students on the Museum of Modern Art field trip got to eat. What ideas about fractions emerge from a comparison of these approaches?** This question asks participants to examine the ideas of equivalency and congruency and to consider how two different fractions can be equivalent without being congruent.

4. **What do Jackie and Ernie need to understand in order to know that the Museum of Modern Art group had less to eat than the Ellis Island group? [lines 165–172]** Ernie and Jackie use what they understand about equivalence when they add $\frac{1}{5}$ (or $\frac{2}{10}$) $+ \frac{1}{10} = \frac{3}{10}$ and $\frac{1}{8} + \frac{1}{4}$ (or $\frac{2}{8}$) $= \frac{3}{8}$.
Furthermore, they know that tenths yield smaller pieces than eighths and that because the numerators in both cases are the same (3), a comparison of the denominators will tell them which group got the most to eat.

In a later chapter in the same Fosnot and Dolk book (Chapter 4, "Developing Big Ideas and Strategies"), the authors explain that when fractions are introduced to students as division within a fair-sharing context, students are less likely to treat them like whole numbers or assume that the greater the number in the denominator, the larger the fraction. Students who have an understanding of fractions as division understand that the opposite is true—the greater the denominator, the smaller the fraction (if the numerators are the same).

Whole-Group Discussion

In this whole-group discussion, participants consider the multifaceted and complex nature of fractions in the elementary curriculum. They also consider the pedagogical choices Ms. Mosseson makes at this early stage of the class investigation of the fair-sharing problem.

♦ **What ideas about fractions surfaced as students worked on this problem?** Many participants may not realize that an understanding of fractions involves comprehension of a number of key concepts, such as part/whole relations, equivalence, and common denominators.

Your group may identify additional mathematical concepts that students in Ms. Mosseson's class are grappling with.

◆ **What is complex about fractions?** Participants may already have touched on this as they worked through the different approaches. As participants discuss this question, ask them for specific examples from the reading to support their ideas.

◆ **What was Ms. Mosseson's approach to working with her students?** Ms. Mosseson poses a rich problem for her students, one that provides them opportunities to encounter some key ideas about fractions. Ms. Mosseson is working to both highlight her students' ideas and gain a deep understanding of how her students are thinking. This lays the groundwork for a later whole-class discussion of the approaches students took to solve the problem.

Because participants do not read the portions of this chapter that deal with how Ms. Mosseson continues to interact with small groups and then conducts a whole-group discussion, you may want to let them know that the point at which the homework reading ends is just the beginning of the class investigation of the fair-sharing situation. Ms. Mosseson continues to provide opportunities to further the development of her students' mathematical thinking. As she moves from group to group, she poses questions that relate to the key mathematical ideas students are exploring in this lesson—part/whole relations, equivalence, common denominators—and she asks them to support and defend their thinking. Following the small-group work, Ms. Mosseson pulls everyone together and chooses groups to present their work. She selects groups whose approaches and strategies are likely to illuminate an understanding of the important mathematical ideas of the lesson and from whose thinking other students would benefit.

You and your participants may be interested in obtaining a copy of *Constructing Fractions, Decimals, and Percents* to read about which mathematical ideas that emerged about the submarine sandwich problem Ms. Mosseson chose to feature in the "Math Congress" she organizes following the small-group work.

ACTIVITY 2
Mathematics Exploration
35 minutes

> ***Materials***
> Handouts 16–19

The Fraction Tracks Game

5 minutes

Introducing the Game Explain to participants that the videotaped classroom episode they will view in this session shows a fifth grade class exploring equivalent fractions by playing the game Fraction Tracks from the *Investigations in Number, Data and Space* unit "Name That Portion." Let participants know that this activity is in the sixth session of the unit, so students have already worked with the Fraction Tracks board and cards and have done tasks such as filling in the missing fractions on the board, finding patterns, and ordering and adding fractions.

Tell participants that they will play Fraction Tracks themselves in order to explore the mathematical ideas involved. Later, they will discuss their own mathematical thinking as they played the game.

Have participants work in groups of four with teams of two competing against each other. Distribute Handouts 16–19 (Directions for Playing Fraction Tracks, Fraction Tracks Board, Fraction Tracks Cards, and Fraction Tracks Reflection Questions).

Inform participants that they will be playing a different version of the game from the one they will see in the classroom episode. They will be playing in teams of two and will compete against another team. Students in the video work together to plan their moves rather than play competitively.

Go over the instructions so that everyone knows how to play the game. Then have pairs fill in the missing dots and fractions along the board so that each track is divided into the number of parts indicated at the far left of the track (halves, thirds, fourths, etc.) and each fraction is labelled.

Ask participants to respond to the questions on Handout 19 as they play the game.

15 minutes

Playing the Game As groups play the game, move from group to group to get a sense of the strategies being used. Take note of which ones you would like to highlight during the whole-group discussion that follows.

> In this discussion participants explore their understanding of equivalent fractions to gain insight into the thinking that some of the children on the videotape may be engaged in as they play the same game.

15 minutes

Discussing the Experience of Doing the Math Have a discussion of the questions on Handout 19.

Sometimes when you picked a card, you may have made an immediate decision about how to use it. Other times you may have spent more time thinking about how to use your card. Give examples of times when these decisions were more complex for you. What made them more complex?

What questions did you ask yourself when making a decision?

What mathematical knowledge were you using in making decisions?

Overview of the Game

This game is played with a Fraction Tracks board, a stack of 20 fraction cards, and a large number of "movers" (chips, pennies, or another kind of marker). Two players, working as a team, compete against another team to move as many chips as they can across the board to land exactly on the number 1. The team that moves the greatest number of chips across the board to 1 is the winner. In this mathematically rich activity, players explore the properties of equivalent fractions.

The Fraction Tracks game, based on a linear model of fractions, helps players understand that fractions that look different, such as $\frac{1}{2}$ and $\frac{2}{4}$, can have the same value; that is, they are equivalent fractions. Another important understanding is that fractions can be taken apart and put back together again in different ways. For example, the fraction $\frac{4}{9}$ can be split into $\frac{1}{9}$ and $\frac{3}{9}$, and $\frac{3}{9}$ is equivalent to $\frac{1}{3}$. As they play the game, players make strategic decisions about how to take apart and recombine the different fractions on the card they draw to move their chips across the Fraction Tracks board in the most efficient manner to score a point.

This activity is taken from the grade 5 *Investigations* unit "Name That Portion" (Investigation 2: Models for Fractions, Session 6). In the unit, students work with clock fractions and fraction strips before they begin to work with the Fraction Tracks game board by filling in the missing fractions on the board, looking for patterns, and counting with and ordering fractions.

The version that participants play differs from the one they will see students playing in that participants play in teams competing against another team. The students work together to plan their moves and do not engage in any competition.

Participants will take a few minutes to fill in the missing fractions before they play the game. Make sure that they understand that they will need to add and label dots to some of the tracks.

Before participants begin playing the game, work through an example that illustrates the kind of strategic thinking that players can use. You may want to make an overhead of the game board to play out the following example:

> You have chips at $\frac{1}{2}$, $\frac{2}{3}$, $\frac{1}{4}$, $\frac{2}{5}$, $\frac{1}{6}$, $\frac{2}{8}$, and $\frac{6}{10}$. You draw a card with $\frac{5}{6}$ on it. What combinations of moves could you make so that more than one chip lands exactly on 1?

For this problem, moving $\frac{5}{6}$ on the sixths line would result in a chip landing on 1. Moving $\frac{1}{2}$ on the halves line and $\frac{1}{3}$ on the thirds line would result in <u>two</u> chips landing on 1.

Discussing the Experience of Doing the Math

After playing the game, participants discuss the mathematical thinking that they engaged in while playing the game. They talk about what made some of their decisions difficult, what questions they asked themselves while making decisions, and what mathematical knowledge they were using in making decisions.

One key idea to bring out is how taking apart and recombining fractions can contribute to a winning strategy. For example, if a player with chips on $\frac{3}{5}$ and $\frac{4}{8}$ draws $\frac{9}{10}$, that player can split the $\frac{9}{10}$ into $\frac{5}{10}$ or $\frac{1}{2}$ and $\frac{4}{10}$ or $\frac{2}{5}$.

The player can move the chips on both the fifths and the eighths lines to 1, earning two points.

This discussion will help participants understand the thinking of the students and concepts students can learn from playing this game. It will give you insights into how the participants in your group think about fractions. This game may stretch the fractions knowledge of some of the participants.

ACTIVITY 3
Video
70 minutes

Materials
Videotaped Clip 4
Handouts 20 & 21
Flip chart or overhead
Markers

Viewing the Fraction Tracks Video

15 minutes

First Viewing of the Videotaped Clip (0:54–14:57) Tell participants that they will now watch the clip in which students explore the same mathematics that participants just did. Give them the following background information about the classroom in the video:

- This is a fifth grade class taught by Ms. Paster.
- Students are exploring equivalent fractions by playing the Fraction Tracks game.
- The Fraction Tracks game is one of a series of games designed to help students consolidate knowledge of equivalency in fractions that they have been exploring in depth prior to this investigation. Thus, it is not meant to be a deep exploration of fractions.
- The 15-minute clip has been slightly edited to show portions of the introduction, small-group work, and whole-group discussion of the investigation. Occasionally one hears Ms. Paster in a voice-over reflection. Despite the edits, participants can get a good feel for the kind of teaching and learning that goes on in this class.
- Students are not playing the game in exactly the same way participants just did.

Participants will use the Overall Observation Guide for this first viewing.

Have participants count off by threes. Group 1 will focus on the mathematics content portion of the guide, Group 2 will concentrate on the learning and pedagogy portion of the guide, and Group 3 will use the intellectual community portion of the guide. Before viewing the video, have participants take a minute to review the questions on their part of the guide to get familiar with what they will be attending to during the first viewing.

After viewing the videotaped clip, give participants a few minutes to jot down notes about what they saw.

15 minutes
Whole group

Discussing the Videotaped Clip Ask for observations from those who used each part of the guide. Ask first for comments from those who looked at the intellectual community part of the guide, then the learning and pedagogy part, and then math content.

Encourage participants to provide evidence from the classroom for the comments they make and to stay close to the mathematics that is happening in the class.

5 minutes

Introducing the Math Content Observation Guide Distribute the Math Content Observation Guide. Explain that, as was the case with the Intellectual Community Observation Guide, the "bulleted" questions are what they should look for in order to develop a response to the header question written in bold.

You might wish to read each of the questions aloud and talk through some of them so that participants get a sense of how this part of the guide works. (See **Facilitator Notes** for a full description of the Math Content Observation Guide.)

Participants may benefit from an example when you go over the questions that refer to the mathematical "topic" and the "ideas within the topic." Mention the importance of looking beyond the general topic being addressed, (e.g., fractions) to what the ideas within this topic that students are exploring (e.g., *When you increase the numerator of a fraction and keep the denominator the same, you have more of that size piece and thus a greater quantity,* or *When you keep the numerator the same and increase the denominator, you have the same number of pieces but the pieces are smaller, and thus you have less.*) Being able to attend to the mathematics at this level of detail can help participants begin to make sense of what it is that students understand and what they are still struggling to understand.

15 minutes

Second Viewing of the Video (0:54–14:57) For the second viewing, all participants will work with the Math Content Observation Guide. Before they view the videotaped clip a second time, suggest that they listen to the students' mathematical talk with an eye to making sense of what they do and do not understand about equivalent fractions. They should try to figure out what mathematical ideas the students are working on.

After viewing the videotape, give participants a few minutes to finish writing their notes about the video.

10 minutes

Small group

Discussing the Videotaped Clip Ask participants to spend 10 minutes in small groups discussing the following questions, modified from those in the Math Content Observation Guide and tailored to this particular video. In thinking about these questions, participants might want to refer to the transcript of the videotape found in the book of *Readings.* (Reading 5)

What specific ideas about equivalent fractions are being explored by different students in this class?

What kinds of mathematical thinking are taking place (conjectures, proofs, revising, generalizing, etc.)?

How does Ms. Paster show that she understands enough about equivalent fractions to test the boundaries of students' understandings?

In what ways does Ms. Paster help the students extend or deepen their thinking?

These questions are particularly appropriate for this episode. They represent the depth to which participants need to understand what is going on in a mathematics class if they are to make well-grounded conclusions about the effectiveness of the class.

10 minutes

Whole-group discussion

Discussing the Video Finish this activity with a whole-group discussion of these questions. Keep track of participants' comments on flip chart paper. Be particularly alert to the issues described in the **Facilitator Notes** for this activity: that participants talk about the range of ideas that students are exploring rather than what students do not understand or how specific Ms. Paster's interventions are.

Overview

This 15-minute videotaped clip shows a fifth grade class exploring equivalent fractions by playing the Fraction Tracks game from *Investigations in Number, Data, and Space.* In previous class sessions, students have worked with the fraction tracks board and cards and have done tasks such as filling in the missing fractions on the board, finding patterns, adding fractions, and putting in order the fraction cards. In this segment the teacher, Ms. Paster, explains how to play the Fraction Tracks game, and then students work collaboratively in pairs. The students are not competing with each other but must work together to decide on the next best move to make. After the students play the game for a few minutes, Ms. Paster pulls them back into whole-group discussion and asks several students to describe an example of what they did in the game.

Discussion of the Videotaped Clip

Intellectual Community

Ms. Paster has taught her students to listen carefully to each other. They know they might have to paraphrase what a classmate said, so they listen to and learn from each other. The involved talk in the class indicates that the students are engaged in mathematical inquiry. Whenever Ms. Paster joins a group, her interactions always focus on students' mathematical thinking. She tunes in to the way they are thinking about mathematics and helps them move forward with their own thinking, asking very specific questions that are grounded in the mathematical ideas that the students are working on. The students show respect for each other's ideas by listening attentively to each other and are able to describe the ideas of their classmates. They use each other as resources by explaining their thinking to each other as they negotiate the best move. The way Ms. Paster has them play this game conveys the message that the thinking of *both* students is needed to figure out the best move.

Usually we see Ms. Paster being a resource to the students. Because of the way the videotape was edited, we rarely see how groups are working when Ms. Paster is not there. This is a significant aspect to observing classrooms. It is important that the observer pay attention to what groups of students do in the absence of the teacher in order to have a full sense of what they can do and how they make sense of mathematical ideas.

Learning and Pedagogy

The students are making sense of how fractions can be decomposed into smaller fractions, what fractions are equivalent, and how fractions can be added to get to 1. (Note that in this game there is no mention of reference wholes. It is assumed that all of the fractions are part of the number 1 and that this reference whole does not change.)

Some of these ideas seem confusing to some students. One student does not understand that $\frac{6}{10}$ is the total on the card and that he needs to identify two fractions that add up to $\frac{6}{10}$. He adds on to $\frac{6}{10}$ instead. Another student re-did an earlier step in order to accommodate the need to use $\frac{9}{10}$, rather than decomposing the $\frac{9}{10}$ itself. It is not clear whether this student does not know how to decompose fractions or whether she simply does not think of the decomposition strategy for this move. Tamara, another student seems to be confused about how to convert one fraction into an equivalent one.

Ms. Paster works with the sense the students are making by creating many occasions for students to express their thinking. She asks questions such as, "How did you know that . . ." or "How was she able to do that?" She also seems to want students to generate many ways to work with any single fraction. She explains how to use two pieces, not just one, in making a move in the game, and she asks them to generate several possible moves on their strategy sheets, encouraging them to develop several equivalencies. Because of the editing of the video, we do not see her working with students who have a fragile understanding, but she clearly has suggested that one girl use the fraction pieces in order to explore what the fractions mean and how they are related. Ms. Paster seems to be attending to *all* students in the class when she says that some students see an easy move and others see more complex moves. It seems that Ms. Paster paired students who tend to engage in complex thinking with students whose ideas tend to be simpler so that the ones who understand more would have the opportunity to explain their thinking to someone else, and the students who are having difficulty understanding would get to see and hear about some different strategies.

Although there is much to appreciate about this video, some participants may raise concerns about aspects of the classroom culture. All the students in this class are working on challenging mathematics, but some participants may note differences in the kind of thinking African American and Caucasian students are shown doing. (There are also Asian children in this class, but we hear from them on one occasion only, and in that case it was in response to a question from Ms. Paster to reiterate what another student has said.) Although African American students are active participants, in several instances they indicate some confusion about the particular concepts brought out in the Fraction Tracks game. It would be important for participants to recognize that we cannot conclude on the basis of this footage that the African American students in this class are less competent in mathematics. Students are complex learners and often their weaknesses are balanced by significant strengths in other areas. If this issue is raised in your group, point out that because we are getting such a brief look at these students there is much about them as learners that we cannot know. Encourage your participants to consider what Ms. Paster might be doing to address the needs of the range of learners in her class. For example, as noted above, her choice of how to group her students may reflect a desire on her part to pair students with a solid understanding of the mathematics of this activity with those whose understanding is not as developed, thinking that both students would benefit from the experience.

A different but related issue may emerge from some participants who may note that at times the Caucasian students are doing most of the thinking and that although Ms. Paster is certainly paying attention to the African American students in her class, in this segment she does not seem to be actively pursuing *their* thinking. One instance participants may point to is toward the beginning of the video when Ms. Paster is interacting with an African American student (whose name we do not learn) and her partner, Sean, who is Caucasian. Ms. Paster seems to be focusing on Sean's ideas only during the entire interaction. Towards the end of their conversation, Ms. Paster suggests to the African American student that she write down Sean's ideas.

If these issues do come up, it would be important for you to facilitate a discussion of them. As we suggest whenever you are discussing videotaped clips with your group, remind them that because we are seeing only a small portion of a class and because the tapes are edited, there is much that we cannot know about what took place. For this reason, it is wise not to jump to conclusions.

In the same way, we cannot make assumptions about the thinking behind the teacher's actions and pedagogical moves. Teaching decisions are very complex and situation-specific, and we simply do not know enough about what was in Ms. Paster's mind as she interacted with her students.

If the issue of equity comes up, you might want to ask your group to think about which elements of this classroom support equal intellectual opportunities for all and what questions related to equity they would want to ask Ms. Paster. Among the supporting elements, they might note the open-ended nature of the activity, which allows for multiple entry points and the careful attention Ms. Paster gives to how her students are thinking about the mathematics. Participants may wish to ask Ms. Paster about how she decided on the groupings she made for her students and where she thought the soft spots in certain students' thinking might be.

Mathematics Content

The students are working on several concepts: that some fractions can be decomposed into smaller fractions; that some fractions are equivalent to other fractions (for example, $\frac{2}{4}$ is equivalent to $\frac{1}{2}$, or $\frac{4}{8}$); and that fractions can be added to make 1. This is worthwhile mathematics because it gives students a great deal of experience thinking about fractions as quantities less than one, making different combinations of fractions, and doing mental math. Ms. Paster clearly understands the mathematics here very well. Her understanding goes well beyond having memorized some rules. She understands that fractions can be decomposed (as illustrated by her example of how to do this when she was explaining how to play the game). She understands that some fractions can be equivalent but expressed in different ways. She also understands that fractions can be expressed as decimals or percents, as illustrated by the chart from a previous lesson on the bulletin board. There is evidence that she has a long-term agenda for student learning because this lesson is part of a series of lessons. Ms. Paster understands that one of the ways students' thinking about fractions may develop is to have them express their thinking aloud and have other students question, challenge, or paraphrase that thinking. She talks quite a bit about this in the voice-over. She also notes that the students expand on the mathematical ideas with their own examples, which indicates that she understands that working *with* ideas in ways that are personally meaningful is one of the ways that children learn.

In pre- and post-observation conferences, it would be important to hear Ms. Paster's thinking about how this lesson fits into a more extended sequence of learning about fractions. For example, Ms. Paster may discuss:

- the range of models that students use for exploring and representing fractions (such as area, linear, time, volume, division, ratio, and so on).
- opportunities they have for exploring the relationship between parts and wholes when working with different quantities.
- ways they come to understand the relationship between fractions, percents, and decimals.

Such information would not only give an observer insight into the teacher's long-term agenda but would also support the teacher in refining that agenda.

Watching the Videotaped Clip and Using the Overall Observation Guide

Participants watch this video the first time while using the Overall Observation Guide. One group of participants observes for intellectual community, one for learning and pedagogy, and the third group for math content. Note that within each category, participants observe both students and teacher, and their comments may be about the *interaction* between teacher and students. The Observation Guide is designed to encourage observation of these interactions and not just the teacher's moves or the students' actions.

Discussion of the First Viewing

One way to facilitate this discussion would be for you to ask for observations about each category in turn and to keep track of participants' comments on flip chart paper. Keep in mind that because there is a great deal of overlap among the categories, it is sometimes difficult to determine into which category a particular comment belongs. When participants present their observations, it is important for you as facilitator to ask them to supply evidence for the comments they are making and to encourage them to stay close to the mathematics that is taking place in the session. If, for example, someone comments that students are building on one another's ideas, ask for an example of where this happens and what the content of the exchange was.

Introducing the Math Content Observation Guide

The Math Content Observation Guide is designed to underscore the central importance of observers attending to the mathematics concepts that are the focus of the lesson and the teacher's knowledge of the mathematics. A focus on the mathematical *ideas* students are working with, the teacher's long-term mathematical agenda, and the teacher's appreciation of the ways students' understandings of mathematical ideas develop over time characterize this approach to mathematics content.

The Questions in the Math Content Guide

The first set of questions on the student side of the Guide **(What is the topic? What are the ideas within the topic that are being explored? What specific ideas are being explored by different students or groups of students?)** is intended to draw the observer's attention to the mathematics concepts that are at play in the activity. Observers are asked to note the general topic being investigated (fractions), as well as the mathematical ideas being explored at a finer level of detail (some fractions are equivalent, fractions can be decomposed into other fractions, and so on).

The second set of questions on the student side **(What important mathematical ideas are involved? What is the relationship between doing procedures and exploring ideas in this mathematics? What kinds of mathematical thinking are happening (conjectures, proofs, revising, generalizing, and so on)?** is intended to encourage observers to look closely at what students are doing from a mathematical perspective. Are they exploring ideas or simply executing a given set of procedures? What kinds of mathematical thinking are they engaged in (making conjectures, developing proofs, revising, generalizing, and so on)?

On the teacher's side, the first set of questions **(What aspects of the mathematics does the teacher have a firm grasp of, and in what areas does she still need to deepen her knowledge? Is the teacher responding to the mathematics in students' mathematical thinking? If so, how? How does the teacher show she understands enough about the mathematics to test the boundaries of students' understandings?)** is intended to draw attention to where the teacher is in terms of his or her understanding of the mathematics and students' thinking.

The second set of questions on the teacher's side **(How does the teacher show that she understands students' thinking? What follow-up questions does the teacher ask to probe the robustness of the students' understanding? In what ways does she help them extend or deepen their thinking?)** underscores the idea that key to making these ideas accessible to students is the teacher's understanding of the mathematics students need to develop and the teacher's capacity to respond to and stretch students' mathematical thinking. A teacher uses a long-term mathematical agenda and a deep understanding of the ways in which students develop ideas about given topics in mathematics to guide decisions about what ideas to follow and build on, which promising insights to highlight, and what potentially "soft" areas to probe for quality of understanding.

The last set of questions on the teacher's side **(What mathematical ideas is the teacher probing? What does the teacher do to bring**

into focus the long-term importance of these mathematical ideas?) calls attention to how the mathematical concepts being explored during the class session fit into the teacher's long-term plan for the students' learning of the mathematics.

Second Viewing of the Classroom Episode

This will be the first of two sessions in which participants focus their attention on the Math Content Observation Guide, although undoubtedly content issues were brought up when participants discussed the nature of the intellectual community in previous sessions of this course. When using the Overall Observation Guide for the first viewing of the *Fraction Tracks* videotape, participants will have undoubtedly noticed a number of content issues. The purpose of viewing the video a second time with the Math Content Observation Guide is to focus on the bulleted items, which draw participants' attention to the elements they should attend to in order to respond to the questions in bold. For example, in order to determine what mathematical ideas students are working with, observers should note:

- the mathematical topic, in this case, fractions or equivalent fractions.
- the ideas or concepts within this topic that are being explored, in this case, that some fractions can be decomposed into smaller fractions, that two equivalent fractions are two different ways of describing the same amount using different-sized parts, and that fractions can be added to make one whole.
- the specific ideas being explored by different students or groups of students; in this case, that equivalent fractions can be represented in different ways, that fractions can be split up to create new equivalent fractions, that fractions have relationships to one another, and that fractions can be manipulated numerically.

Although some of the bulleted items will have been addressed during the first viewing of the episode, the second viewing is an opportunity for participants to do a more detailed observation using only the Math Content Observation Guide.

The discussion questions for the second viewing are iterations of the questions and bulleted items on the Content Guide made specific for this particular videotaped clip, plus a few other questions that are particularly appropriate. The questions are listed below, with examples of the kinds of answers participants might give.

Small-Group Discussions

♦ **What specific ideas about equivalent fractions are being explored by different students in this class?** The wording of this question is important. It does not ask which students are having trouble and what they do not understand. It asks what *ideas* are being explored by many different students, including those students whose understanding is fragile. In every class students are at very different places in their thinking about a given mathematical idea, and teachers have to make instructional decisions on the basis of that reality. In responding to this question, participants need to adopt a listening attitude to understand what ideas various students are working on.

Here are some of the ideas about fractions that the students in Ms. Paster's class are working on. Participants may identify others.

1. One girl at the beginning says that one could put either $\frac{8}{10}$ or $\frac{4}{5}$ on the board because they are equivalent. She is exploring the idea of multiple representations of the same quantity.
2. Jo uses the fraction pieces to think about how the fractions relate to one another and which ones are equivalent. For example, she

puts nine $\frac{1}{10}$ pieces on a black rectangle in order to see what $\frac{9}{10}$ is. Jo seems to be working on visualizing how fractions relate to one another and to 1. She does not yet seem to be working on the idea that one can take fractions apart and recombine them.

3. Several children in different groups are working on decomposing fractions and identifying new equivalent fractions so that they can "jump" from one track to another. For example, one boy says that he and his partner Tamara were stuck because they could not tie sixths and eighths together. They had $\frac{2}{8}$ on the board, and drew the card $\frac{8}{8}$. They knew that they could go from $\frac{2}{8}$ to $\frac{8}{8}$, using $\frac{6}{8}$, with $\frac{2}{8}$ left over. He and Tamara were working out the idea that they could move a chip to $\frac{1}{4}$ and that would use up the the $\frac{8}{8}$ on the card. While we watched, the boy discovered that $\frac{1}{4} = \frac{2}{8}$ and therefore $\frac{6}{8} + \frac{1}{4} = \frac{8}{8}$.
4. Tamara says that she started with a chip on $\frac{2}{6}$ and drew $\frac{6}{6}$, so she counted 4 up from $\frac{2}{6}$ and got $\frac{6}{6}$, which was one. She then tried to use what was left over, $\frac{2}{6}$. She says that half of 6 is 3, so she went to the $\frac{1}{3}$ line. She then counted 5, 6, which got her to the $\frac{1}{3}$ mark. She said that she knew that $\frac{1}{3}$ is $\frac{2}{6}$ (which is the amount she had left over) and that the $\frac{1}{3}$ would be half of the $\frac{2}{6}$. It seems that Tamara is trying to change the $\frac{2}{6}$ into $\frac{1}{3}$, which is a legitimate move in Fraction Tracks because they are equivalent fractions and she may be doing this visually by looking at what is exactly above $\frac{2}{6}$ on the Fraction Tracks chart. But her explanation about why this works is not valid. Tamara still seems to be working on how to convert one fraction to another and is using a procedure that works for whole numbers but doesn't work for fractions.

♦ **What kinds of mathematical thinking are taking place (conjectures, proofs, revising, generalizing, and so on)?** Most of these students are describing their strategies—for example, describing what calculations they made to figure out a Fraction Tracks move—or are paraphrasing other students' ideas. In the process of developing their Fraction Tracks moves, they have to provide mathematical justifications for their moves (for example, "$\frac{2}{8}$ is equal to $\frac{1}{4}$, so I can move a chip on the $\frac{1}{4}$ line.")

♦ **How does Ms. Paster show that she understands enough about equivalent fractions to test the boundaries of students' understandings?** We do not see Ms. Paster doing much of this. However, in one instance Ms. Paster is working with two students, Tamara and a boy whose name we do not learn, exploring what to do with an $\frac{8}{8}$ card when the chip is already at $\frac{2}{8}$. The boy reasons that you could use $\frac{6}{8}$ to get to 1 then and use the remaining $\frac{2}{8}$ to go to $\frac{1}{4}$. Tamara thinks that she can move a chip up to $\frac{2}{4}$ because if you have used $\frac{6}{8}$, you need to use 2 more eighths to reach $\frac{8}{8}$, and you have $\frac{2}{8}$ left over. Tamara then wants to go to $\frac{2}{4}$, not converting $\frac{2}{8}$ into $\frac{1}{4}$. As she works with this small group on the details of how to work with eighths, Ms. Paster discovers the boundaries of Tamara's thinking. Tamara knows how to add the fractions together to reach 1, but she doesn't yet know how to convert the leftover $\frac{2}{8}$ into fourths.

♦ **In what ways does Ms. Paster help the students extend or deepen their thinking?** At the beginning of the video, when Ms. Paster suggests that the students can make *two* moves for each fraction, she pushes their thinking about how they might create equivalent fractions. Rather than simply identify the equivalent fraction for a number $\left(\frac{8}{10} = \frac{4}{5}\right)$ as the first student did, she suggests that they can decompose the fraction into *two* pieces $\left(\frac{4}{10} + \frac{2}{5} = \frac{8}{10}\right)$.

She notes at the beginning that students might find more than one way to move their pieces across the board. She suggests that they generate several possibilities, enter them on the Fraction Tracks Strategy Chart, and then think about which would be the best move.

She suggests that Jo explore the fractions using the Fraction Pieces so that Jo can see how the fractions are composed and how they relate to one another. Ms. Paster then asks, "What are some of the things we can do with $\frac{9}{10}$?" suggesting that they use the strategy with the Fraction Pieces to explore a different fraction.

She has the students explain their strategies to one another because she believes that when students "bring things to their own language, they expand upon the ideas with their own examples, and they question and challenge each other."

ACTIVITY 4
Homework Discussion
15 minutes

Materials
Handout 14
Reading 6

Discussion of *Magical Hopes: Manipulatives and the Reform of Math Education*

15 minutes
Whole-group discussion

Discussing Manipulatives Remind participants that as homework for this session they read Deborah Ball's article, *Magical Hopes: Manipulatives and the Reform of Math Education* (Reading 6). They were to identify three ideas in the article that interested them and write at least a paragraph about each. Ask for a volunteer to name an idea that interested him or her and explain why. See whether there is more discussion on that point. Then ask for a different idea and have a discussion on that point. You should be able to discuss three or four different ideas.

Toward the end of this discussion ask, the following questions:

In the videotaped clip we just watched, what mathematical understanding was made possible by Ms. Paster's decision to use the Fraction Tracks game as a context for exploring equivalent fractions?

What might have been a disadvantage of using this tool in this situation?

Discussion of *Magical Hopes*

In her article *Magical Hopes: Manipulatives and the Reform of Math Education,* Deborah Ball makes the distinction between the mathematical ideas that may be under discussion in a classroom and the medium through which those ideas are represented. She calls these various media "tools" for exploring, representing, and communicating mathematical ideas. Among these tools are concrete models and materials, graphs and pictures, calculators and computers, and nonstandard and conventional notations. Ball insists that manipulatives, or concrete objects, are important but no more so than other tools in supporting mathematics teaching and learning.

In fact, a major choice that teachers make is the selection of which set of tools to use when working with children on a particular idea. Sometimes, as Ball indicates, none of these tools is necessary. Instead, mathematical discussion of ideas in the classroom is sufficient. At other times, the teacher may wish to use a tool. In that case, the tool needs to represent well the particular aspect of the mathematical idea that the teacher wants children to explore.

In the videotaped clips shown in this course, teachers have made different choices about the use of tools. Sometimes those have been good choices, but at other times the mathematical ideas and the choice of tool do not seem well matched.

This idea may be clear to many participants in your course, but others may still equate standards-based mathematics instruction with the use of manipulatives. Therefore, we assigned this article to give you the opportunity to explore this issue with your group. The assignment was to identify three ideas in this article that participants found interesting and write at least a paragraph about each. You may wish to collect this homework as a way of getting a picture of where the thinking of participants is on this issue.

When you discuss this reading, you will want to hear all views, but try to encourage the development of an analytical orientation toward the use of manipulatives or other tools of instruction. A judgment about the use of the tool depends on the adequacy of the tool for making transparent the mathematical ideas under discussion and the teacher's purposes. The simple presence of manipulatives does not necessarily indicate a good mathematics class.

The final discussion questions help with the development of this analytical orientation.

- **In the videotaped clip we just watched, what mathematical understanding was made possible by Ms. Paster's decision to use the Fraction Tracks game as a context for exploring equivalent fractions?** This game helped support the understanding that some fractions can be decomposed into smaller fractions, that equivalent fractions are different ways of describing the same amount using different-sized parts, and that fractions can be added to make one whole.

- **What might have been a disadvantage of using this tool in this situation?** Manipulatives can work to limit understanding of a concept. One example of this in the Fraction Tracks situation is that students might get the idea that all fractions have the same reference whole.

CLOSING
20 minutes

Materials
Handouts 22, 23, 24

Homework

10 minutes

Assigning Homework Let participants know that there are three homework assignments for this session.

1. Read the article *Capturing Teachers' Generative Change: A Follow-up Study of Professional Development in Mathematics* by Franke *et al.* (Reading 7) and answer questions. Let participants know that this article will introduce them to the idea of "generative learning," which will thread its way through the rest of the course. This idea is part of the second strand of the course, *Rethinking the ways we talk to teachers.*
2. Use the Math Content Observation Guide as they carry out a second classroom observation.
3. Think about the nature of their own communication with teachers when they conduct a post-observation conference. Participants reflect in writing about a post-observation interview they conducted recently and then make a pie chart of the modes of communication that they employed. Distribute to each participant a sheet of chart paper on which they will draw their pie chart. Suggest that they first make a pie chart on the blank pie chart in the book of *Readings* and then copy it onto the chart paper.

Bridging to Practice

10 minutes

Reflective Writing Remind participants that Bridging to Practice is designed to allow them to focus on the ideas in the course that seem particularly interesting for them and for their school community. Point out that it provides a framework for planning what they can do in order to move themselves and their schools along with these ideas.

Post the following questions, and invite participants to respond to them:

Choose an idea that came up today that you found particularly interesting. What is your current thinking about this idea?

Where is your school now with regard to this idea?

What are one or two things that you, as instructional leader, will pursue to move yourself and/or your school along with this idea?

Your Own Journal Writing

Within one or two days of teaching this class, after you have had the chance to unwind from the class and perhaps talk with a colleague or friend about how it went, set aside some time to write your own journal entry about the class. We encourage you to write about the ideas held by the participants in your group rather than your own actions and sense of how things worked out. In our experience, writing about what the participants are thinking about will be much more useful to you as you prepare for the next session. Some facilitators use the time when administrators are writing in their own journals to make preliminary notes for this journal writing.

Considering the Complexity of Fractions

1. When Ms. Mosseson asks her students whether all groups had the same amount to eat, some of her students respond that the students in the three groups where there was one fewer submarine sandwich than students would each have the same amount to eat [lines 29–47]. What seems to be going on for the students who respond this way?

2. What are Ernie and Jackie puzzling about when they are trying to determine how much the Museum of Modern Art group got to eat? [lines 106–132]

3. Compare the strategies that Jackie and Ernie on the one hand [lines 106–132] and Nicole and Michelle on the other [lines 74–86] use to figure out how much the students on the Museum of Modern Art field trip got to eat. What ideas about fractions emerge from a comparison of these approaches?

4. What do Jackie and Ernie need to understand in order to know that the Museum of Modern Art group had less to eat than the Ellis Island group? [lines 165–172]

Directions for Playing Fraction Tracks[1]

This game offers players the opportunity to explore the properties of equivalent fractions and appreciate how fractions can be combined and taken apart. Two players, working as a team, compete against another team to move as many chips as they can across the board to land exactly on the number 1. The team that moves the greatest number of chips across the board to land on 1 is the winner.

To begin the game: Teams share a game board. Place seven chips on a game board, one on each track, at any fraction point less than $\frac{3}{4}$. Mix the cards, and place the deck face down. Each team, in turn, draws a card from the deck and moves a chip or a combination of chips along from one fraction to the next on the game board for a total move equal to the fraction on the card.

Teams must move only the exact fraction on their card at each turn and they must use up the whole fraction. A team may not move an individual chip beyond the 1 or "wrap it around" to restart the chip at the beginning of the same track during a turn.

Chips are left in position on the board at the end of each team's turn and become starting points for the next team's turn.

Occasionally, a team may draw a card that they cannot use. In this case, it is the other team's turn.

When a team lands a chip or chips on 1, the team collects the chip(s) and puts a new chip on the same track at 0.

Game Strategies: Expressing fractions or taking them apart in different ways often makes it possible to move chips on two or more different tracks to 1 during the same turn, thus earning more than one point on a turn. For example, if the card drawn is $\frac{3}{5}$, a chip can be moved $\frac{3}{5}$ on the fifths line, or $\frac{6}{10}$ on the tenths line, or a combination of moves on two or more lines, such as $\frac{1}{2}$ and $\frac{1}{10}$; or $\frac{1}{5}$ and $\frac{4}{10}$; or $\frac{1}{3}$, $\frac{1}{6}$, and $\frac{1}{10}$.

[1] Adapted from *Investigations in Number, Data and Space: Name That Portion (Fractions, Percents and Decimals)*, Grade 5. "Fraction Tracks" ©1996 Dale Seymour Publications. Used by permission of Scott Foresman and Company.

Fraction Tracks Board

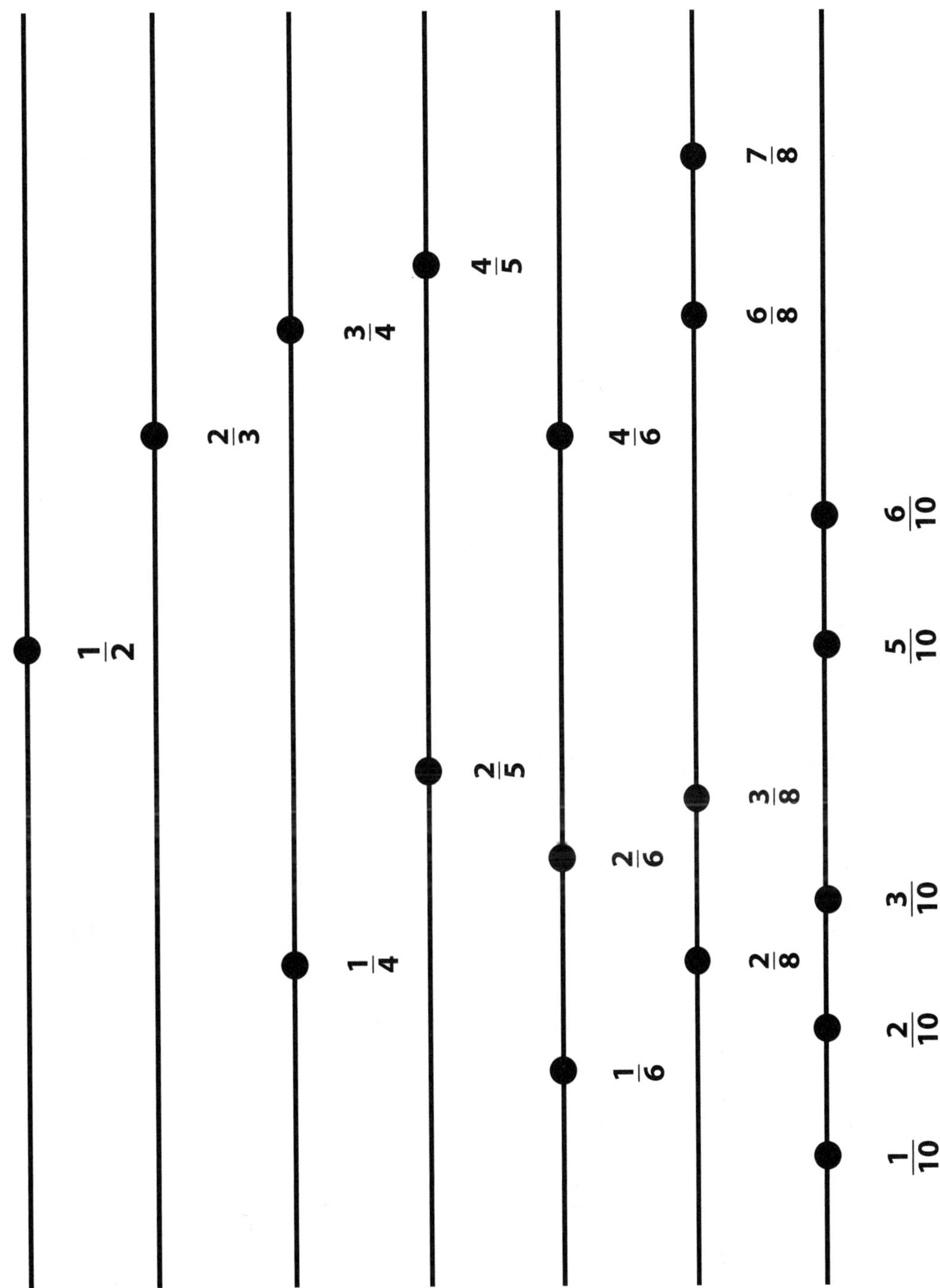

Adapted from *Investigations in Number, Data and Space: Name That Portion (Fractions, Percents and Decimals)*, Grade 5. "Fraction Tracks" ©1996 Dale Seymour Publications. Used by permission of Scott Foresman and Company.

Fraction Cards

$\frac{1}{2}$	$\frac{1}{3}$	$\frac{2}{3}$	$\frac{1}{4}$
$\frac{3}{4}$	$\frac{1}{5}$	$\frac{2}{5}$	$\frac{3}{5}$
$\frac{4}{5}$	$\frac{1}{6}$	$\frac{5}{6}$	$\frac{1}{8}$
$\frac{3}{8}$	$\frac{5}{8}$	$\frac{7}{8}$	$\frac{1}{10}$
$\frac{3}{10}$	$\frac{7}{10}$	$\frac{9}{10}$	$\frac{2}{2}$

Adapted from *Investigations in Number, Data and Space: Name That Portion (Fractions, Percents and Decimals)*, Grade 5. "Fraction Tracks" ©1996 Dale Seymour Publications. Used by permission of Scott Foresman and Company.

$\frac{2}{4}$	$\frac{4}{4}$	$\frac{5}{5}$	$\frac{2}{6}$
$\frac{3}{6}$	$\frac{4}{6}$	$\frac{2}{8}$	$\frac{4}{8}$
$\frac{6}{8}$	$\frac{2}{10}$	$\frac{4}{10}$	$\frac{5}{10}$
$\frac{6}{10}$	$\frac{8}{10}$	$\frac{10}{10}$	$\frac{1}{1}$

Adapted from *Investigations in Number, Data and Space: Name That Portion (Fractions, Percents and Decimals)*, Grade 5. "Fraction Tracks" ©1996 Dale Seymour Publications. Used by permission of Scott Foresman and Company.

Fraction Tracks Reflection Questions

As you play the Fraction Tracks game, jot down your thoughts about the following questions:

1. Sometimes when you picked a card you may have made an immediate decision about how to use it. Other times you may have spent some time thinking about how to use your card. Give examples of times when these decisions were more complex for you. What made them more complex?

2. What questions did you ask yourself when making a decision?

3. What mathematical knowledge were you using in making decisions?

Observation Guide

Students		
Focus Question	Conjectures	Evidence from Classroom
Math Content		
• What mathematical ideas are embedded in the lesson?		
• What makes this worthwhile mathematics?		
Learning		
• What kinds of mathematics sense-making are students doing?		
• What mathematical ideas seem to be confusing to students?		
• In what ways can you see that the students are developing their mathematical ideas over time?		
Intellectual Community		
• How are students showing respect for one another's ideas?		
• How do students use each other as resources as they make sense of mathematical ideas?		
• What evidence beyond raised hands do you have that students are engaged?		

Observation Guide

Teachers		
Evidence from Classroom	Conjectures	Focus Question
		Knowledge of Content
		• What does the teacher seem to understand about the mathematics?
		• What is the teacher's long-term mathematical agenda?
		• What does the teacher seem to understand about the development of children's ideas in this topic?
		Pedagogy
		• How does the teacher work with the sense the children are making?
		• How does the teacher work productively with students' confusion?
		• How does the teacher attend to all students?
		• How does the teacher adjust her teaching based on the ideas she hears from students?
		Facilitating Intellectual Community
		• How does the teacher support students in showing respect for one another's ideas?
		• How does the teacher set the tone so students see each other as resources for mathematical thinking?
		• What interventions does the teacher make to ensure that students' engagement has a focus on mathematical ideas?

Math Content Observation Guide

Students		
Focus Question	Conjectures	Evidence from Classroom
What mathematical ideas are embedded in the lesson?		
• What is the topic?		
• What are the ideas within this topic that are being explored?		
• What specific ideas are being explored by different students or groups of students?		
What makes this worthwhile mathematics?		
• What important mathematical ideas are involved?		
• What is the relationship between doing procedures and exploring ideas in this mathematics?		
• What kinds of mathematical thinking are taking place (conjectures, proofs, revising, generalizing, etc)?		

Math Content Observation Guide

Teachers		
Evidence from Classroom	Conjectures	Focus Question
		What does the teacher seem to understand about the mathematics?
		• What aspects of the mathematics does the teacher seem to know well and in what areas does he or she still seem to need to deepen her knowledge?
		• Is the teacher responding to the mathematics in students' mathematical thinking?
		• How does the teacher show that he or she understands enough about the mathematics to test the boundaries of students' understanding?
		What does the teacher seem to understand about the development of children's ideas in this topic?
		• How does the teacher show that he or she understands students' thinking?
		• What follow-up questions does the teacher ask to probe the robustness of students' understanding?
		• In what ways does the teacher help them extend or deepen their thinking?
		What seems to be the teacher's long-term mathematical agenda?
		• What mathematical ideas is the teacher probing?
		• What does the teacher do to bring into focus the long-term importance of these mathematical ideas?

Homework for Session 5

A. What Is Generative Learning?

Read *Capturing Teachers' Generative Change: A Follow-up Study of Professional Development in Mathematics* by Megan Loef Franke, Thomas P. Carpenter, Linda Levi, and Elizabeth Fennema (Reading 7).

Skim pages 77–82, and then skip to page 87 and read more carefully.

Two notes:

- The idea of generative learning is expressed throughout this paper in several related terms: "generative growth," "generative change," and "generativity."
- This paper was written for a research audience rather than a practitioner one, so you may find it more technical than what you usually read.

Jot down some thoughts on the following questions:

1. How do the authors characterize generative learning?

2. What view of teachers-as-learners do the authors articulate?

3. What do teachers need to know to be generative learners?

4. Why does a focus on children's mathematical thinking help promote generative learning in teachers?

B. Observing for Content

Do an observation of one of the participating teachers in your school or district. Use the attached Pre-observation Conference Questions to help you prepare for the observation. Use the Math Content Observation Guide for these observations.

Write up a short summary of your observation, addressing as many of the questions in the Math Content Observation Guide as are relevant. Conclude with any questions or issues you are puzzled about.

C. Talk with Teachers

In the next class session we will focus on the "voices" we use in communicating with teachers about what we observe in their classrooms.

For this assignment you will need to have in mind a post-observation conference that you have conducted with a teacher. You can plan to conduct one with the teacher you have just observed or consider one that you did recently.

Reflect in writing on the kind of conversation that took place. Think about the nature of both your and the teacher's role in the conversation. Pay particular attention to the mode of communication that you employed. To what extent were you:

- giving advice?
- criticizing?
- asking questions and finding out what the teacher intended?
- sharing curiosity about students' thinking?
- problem-solving collaboratively on specific issues?
- praising or affirming?
- other?

Fill in the outline pie chart to show the proportions of each of these modes that you used in your end of the discussion. Be sure to use the color key on the chart. Copy your chart onto the large chart paper that the facilitator will give you.

Modes of Communication

Color Key	
Red:	Giving advice
Blue:	Criticizing
Yellow:	Asking questions and finding out what the teacher intended
Orange:	Sharing curiosity about students' thinking
Purple:	Collaborative problem-solving on specific issues
Green:	Praising or affirming
White:	Other

Pre-observation Conference Questions

In order to help you make sense of what you will be seeing when you do your classroom observations, plan to meet with the teacher prior to the observation and ask the following questions:

1. What topic will you and your students be working on in this lesson?

2. What do you plan to do in this lesson? (e.g., the origin and structure of the lesson, and so on)

3. What do you hope to accomplish in this lesson?

4. What mathematical ideas are embedded in this lesson?

5. What have you and your students been working on prior to this lesson?

6. How does this lesson fit into your overall goals for the year?

7. Are there students who have special issues in the class?

SESSION
5

Supporting Generative Learning: How We Talk with Teachers

> Generativity refers to individuals' abilities to continue to add to their understanding. . . . If teachers can learn to talk to their students about their thinking, puzzle about what the responses tell them about students' understanding, decide how to use this knowledge in planning instruction and interacting with students, and figure out how to learn more about the students' thinking—the teachers' own learning can become generative.[1]

In this session, we continue our exploration of the first strand in this course, developing an eye for mathematics classrooms. Participants use the Math Content Observation Guide to focus on the mathematics content of a first grade mathematics lesson. In addition, we begin explicit consideration of the second strand of the course, rethinking the ways we talk with teachers about what we see. These two strands will continue throughout the remainder of the course.

Participants continue to think about what is involved in developing a focus on the mathematics when observing in a classroom. In addition, participants begin to think about how they talk to and with teachers. They explore how the orientation they take when talking with teachers can help teachers develop a more robust understanding of their students' mathematical thinking. Participants also think through why such an orientation is important.

This session invites participants to explore the idea of "generative learning," the process of continually adding to what one understands, and the supervisory practices that best support it. When individuals are engaged in generative learning, they are growing and refining their knowledge. Participants will consider how supervisors and teachers can together engage in generative learning. They explore "collaborative inquiry" (or "co-inquiry"), in which teacher and administrator share their curiosity and their questions about the student thinking that occurred in the observed class and probe more deeply into that thinking.

---

[1]From Franke, M. L., Carpenter, T. P., Levi, L., & Fennema, E. (2001) Capturing teachers generative change: A follow-up study of professional development in mathematics. *American Educational Research Journal, 38*(3), 653–689

Overview for Session 5

OPENING page 172 5 minutes	**INTRODUCTION** This is a time for announcements and an introduction to the session.
ACTIVITY 1 Homework Discussion page 174 35 minutes	**SHARING OF PIE CHARTS** Participants use their analyses of a recent post-observation conference to examine the modes of communication they employed. They think about categories of communication and their impact on teacher learning. Participants will be introduced to the concept of collaborative inquiry (or co-inquiry) as one way of talking with teachers after an observation.
ACTIVITY 2 Video page 179 65 minutes	**FOCUSING ON THE MATHEMATICS CONTENT OF CLASSES** Participants discuss the content observations that they did for homework and view and discuss a videotaped clip of a second grade class exploring geometric shapes. Before viewing the video, participants will do a short mathematics investigation, using pattern blocks to explore fractional relationships.
ACTIVITY 3 Discussion of Reading page 185 20 minutes	**WHAT IS GENERATIVE LEARNING?** Participants discuss the article read for homework. Participants consider what it means for teachers to engage in generative learning and the view of "teachers-as-learners" that underlies the idea of generative learning.
ACTIVITY 4 Role Play and Discussion page 188 40 minutes	**TAKING DIFFERENT ORIENTATIONS TOWARD TALKING WITH TEACHERS** Participants experiment with different kinds of post-observation discourse styles as they engage in role play of a post-observation conference. They consider what the effects of different orientations might be on supporting teachers as generative learners.
CLOSING page 193 15 minutes	**HOMEWORK** There are two homework assignments for this session. Participants will do the mathematics that will be on the videotaped clip for Session 6, and they will reread *What's All This Talk About "Discourse"?* by Deborah Ball. (Reading 3) **BRIDGING TO PRACTICE** Participants finish the session with journal writing that allows them to develop further some of their thinking from the session.

BIG IDEAS

Participants explore:

- attending to mathematical content when doing classroom observations
- teachers as constructors of knowledge about the mathematical thinking of the children in their classroom
- collaborative inquiry between administrators and teachers as a support for teachers' generative learning

MATERIALS

- ☐ nametags
- ☐ flip charts or overheads
- ☐ markers
- ☐ agenda
- ☐ Handout 22, Homework for Session 5
- ☐ Handout 23, Modes of Communication Pie Charts
- ☐ Reading 7, *Capturing Teachers' Generative Change: A Follow-up Study of Professional Development in Mathematics* by Franke *et al.*
- ☐ Videotaped Clip 5, *Pattern Blocks* (DVD #1; Program 8; 16:08–28:50)
- ☐ Pattern blocks: hexagons (yellow), rhombuses (blue), trapezoids (red), and triangles (green)
- ☐ Handout 25, Taking Different Orientations
- ☐ Handout 26, Reflection on Role Play (2 per participant)
- ☐ Handout 27, Math Content Observation Guide
- ☐ Handout 28, Homework for Session 6
- ☐ Reading 3, *What's All This Talk About "Discourse"?* by Deborah Ball

PREPARATION

- Prepare an agenda for the session on an overhead or on flip chart paper with approximate times for each activity noted.
- Write the prompts for Bridging to Practice on an overhead or on flip chart paper.
- As you look over the discussion questions for the activities in this session, think about which ones your group would most benefit from considering and use these to guide your discussions.
- Preview Videotaped Clip 5, *Pattern Blocks,* for Activity 2. Watch it several times: once just to see what it is about, once using the Overall Observation Guide, and once using the Math Content Observation Guide. Because this session is so full, however, participants will view this classroom episode only once.
- Do the mathematics challenge to try to anticipate questions that may come up in Activity 2. Make sure that you have found all of the ways of covering the hexagon. Think about the fractional relationships among the pieces.
- Read thoroughly *Capturing Teachers' Generative Change: A Follow-up Study of Professional Development in Mathematics* by Franke *et al.* (Reading 7) for Activity 3. Review the questions, and anticipate the topics participants will find noteworthy.
- Read through the three scenarios that participants will be using in the role play in Activity 4, and take note of the differences among them. Work through the discussion questions yourself.

OPENING
5 minutes

Materials
Name tags
Agendas

Introduction

Before starting the session, post the agenda in a place where everyone can see it.

5 minutes

Introducing the Session Remind participants that both Session 4 and this session draw their attention to the mathematics content of classrooms as they reorient their focus in observations. Mention that in this class they will continue to focus on the gradual development of students' mathematical ideas and the range of mathematical ideas that students in any class will be working on.

Mention to participants that during the first four sessions of the course, they have focused on *what* to talk to teachers about in conjunction with the classroom observations they conduct. From this point on they will also be thinking about *how* to talk to teachers and what orientation to take in both the pre- and post-observation conferences.

Let participants know that later in this session they will explore the idea of generative learning that they read about for homework. They will consider how generative learning might guide the supervision work that they do with teachers.

Highlight the idea of "collaborative inquiry," or "co-inquiry," in which teacher and administrator share their curiosity and their questions about the student thinking that was observed in the class and collaboratively probe more deeply about that thinking. Frame it as a useful idea to think about in the context of supervisory practices that support generative learning.

Facilitator's Notes
Opening

Overview

This session marks the halfway point in the course. You should know your participants quite well by this time and be aware of the continuing issues that come up for them. By this time, you can use their journal writing to see patterns in their thinking about the central ideas of the course.

In this session, participants continue to focus on the mathematical content in classrooms. They are introduced to the notion of generative learning and explore talking with teachers from a stance of collaborative inquiry.

The **Facilitator Notes** give detailed background about the mathematics in the video in order to prepare you to facilitate an exploration of the ideas in the lesson. They also provide background on "collaborative inquiry" and the central role it can play in classroom observation and teacher supervision. The Notes also discuss the idea of generative learning and highlight its relevance in the context of classroom observation and teacher supervision. Again, the Notes are meant to give you insights into the key ideas of the session and prepare you to facilitate discussions that can go beyond the initial responses to the discussion questions.

ACTIVITY 1
Homework Discussion
35 minutes

Materials
Handout 23

Sharing of Pie Charts

5 minutes

Making Personal Charts Public Ask participants to post around the room the pie charts they drew for homework.

Next invite participants to look around the room at all of the pie charts and reflect on what they see. You might ask:

What do you notice as you look at all of the pie charts around the room?

15 minutes
Whole group

Discussing Individual Charts After looking at characteristics that are similar across a number of the pie charts, focus on a single chart at a time and have the person who made it tell a little about the situation. You might choose a chart that is typical of a subset of charts or unusual in some way and ask the author of the chart to set the context the conversation.

> The purpose of this activity is to make participants conscious of their discourse patterns when conducting post-observation conferences with teachers. Then they can think about what they and the teachers gain and/or lose from their approach to conferencing.

The following are questions you might ask:

What goals did you have for this conversation?

How did you make your decisions on how to approach the conversation?

How did your decisions influence the conversation and the teacher's reaction?

Reflecting back on the conversation, what do you think?

15 minutes
Whole group

Discussing the Charts as a Whole From these individual cases, return to the general, looking at different "profiles" that emerged, the resulting responses (e.g., what happened when supervisors spent most of their time giving advice, trying to understand the teacher's thinking and intentions, sharing curiosity about students' thinking, and so on and what made supervisors decide that the approach to communication they used was the most appropriate one in this circumstance). You could ask:

Are there several different kinds of profiles represented here?

From what circumstances did these profiles arise, and what was the result?

If there is time, invite each participant to make a new pie chart that reflects his or her current perspective on which modes of communication would best match the needs of the teacher he or she was meeting with.

Explain to participants that a later activity in this session will give them the opportunity to practice using different communication approaches when role-playing a post-observation conference.

Overview

Increasingly, there is recognition that teachers need to be much more than enactors of ideas articulated in a curriculum or in a given body of research. Indeed, teachers need to construct their own meaning of new ideas about mathematics, teaching, and learning; make sense of them with reference to the students with whom they work; and cultivate the judgments that support them in making productive instructional choices in their day-to-day practice in the classroom.

Participants examine several different communication modes that they might use when talking with teachers after an observation, with an eye to assessing the impact of each on teacher learning. These modes include co-inquiry or collaborative inquiry, an idea that relates to Nolan and Francis' view that new ideas about learning affect not only how we teach students but also how supervisors might "teach" teachers. Nolan and Francis suggest that just as teachers work with students to help them construct their knowledge of mathematics, supervisors can collaborate with teachers to help them construct their knowledge of students' mathematical thinking and mathematics teaching.

In this activity, participants think about what supervisory practices might best support teachers' learning. Using the pie charts they made for homework, participants discuss the modes of communication they used and their effect on teacher learning. They consider the idea of collaborative inquiry, in which teacher and administrator share their curiosity and questions about student thinking that occurred in the observed class and collaboratively inquire more deeply about that thinking. This activity is designed to help participants rethink both what is fruitful to discuss in post-observation conferences and how such conversations might best support teachers' (and administrators') learning. This homework assignment is part of the process of rethinking the *how* of talking with teachers.

For a second homework assignment, participants chose a post-observation conference that they had recently conducted with a teacher and were asked to:

1. reflect in writing on the conversation that took place and the nature of both the teacher's and their own role in it.
2. make a pie chart of the proportions of each of the modes of communication that they used.

Administrators may not have had much experience considering how their modes of communicating with teachers and with colleagues could be analyzed and categorized for different communication styles. Further, the idea that they might learn something about themselves and their own assumptions about supervision by looking at their communication styles may be fairly new to them.

This activity is grounded in pedagogical assumptions that guide standards-based efforts more generally, one of which is the following: just as teachers have been exploring ways to open up their own teaching to include the students' voices, administrators are encouraged to consider different ways to interact with teachers so that they can hear the teachers' thinking about their students' mathematical thinking. We have chosen six different communication modes, which are based on the categories that have been salient when we have done this activity with other administrators. These categories are **giving advice, criticizing, asking questions and finding out what the teacher intended, sharing curiosity about children's thinking, solving problems collaboratively, praising or affirming, and other.**

When we taught this course a few years ago, the **giving advice** category was dominant in many of the charts. Some administrators indicated that criticizing also represented a significant portion of their charts, in many cases more than they thought it would. The category of **solving problems collaboratively** was also well represented. This is of interest because this behavior, in which principals engage quite often, is a precursor to co-inquiry in the sense that the administrator and the teacher are collaborating to determine what would be a good next step for the teacher to take. The category **sharing curiosity about children's thinking** was not as well represented. Some administrators even noted that they never really considered using such a mode of communication when conferencing with a teacher.

The pie chart representation of their modes of communication can be a very powerful stimulus for raising awareness about one's communication style and suggesting possibilities for a different kind of learning that could be part of a supervisory relationship.

In a more recent offering of the course, the responses were more evenly spread among the choices. Nearly every participating administrator had responses that were categorized as *inquiry*. (See Figure 1.) We speculate that in the interval between the two courses, important ideas about standard-based mathematics education had become more widespread and administrators had begun to integrate them into their practice.

After teaching this course a number of times, we adjusted the different categories. We dropped the **consoling** category because few administrators listed that in their charts. The category of **solving problems collaboratively** was added in response to the frequency that participants in earlier courses included it in the **other** category. (Interestingly, these administrators seemed very conscious of *not* giving too much advice during their conferences and seemed almost giddy when the role plays in a later activity allowed them to freely dispense advice!)

As a whole, later groups of administrators had more inquiry-style comments than earlier groups; still, a number of administrators thought that they needed to work on integrating questions about students' mathematical thinking into their post-observation conversations. They were actively thinking about how this mode of communication would help the teachers redirect their attention toward students' mathematical thinking. Administrators said:

> Looking at this pie chart makes me think I need to do more with students' thinking because I think that's how we need to start looking at evaluation now, not so much in terms of teacher behavior, but in terms of children's thinking.

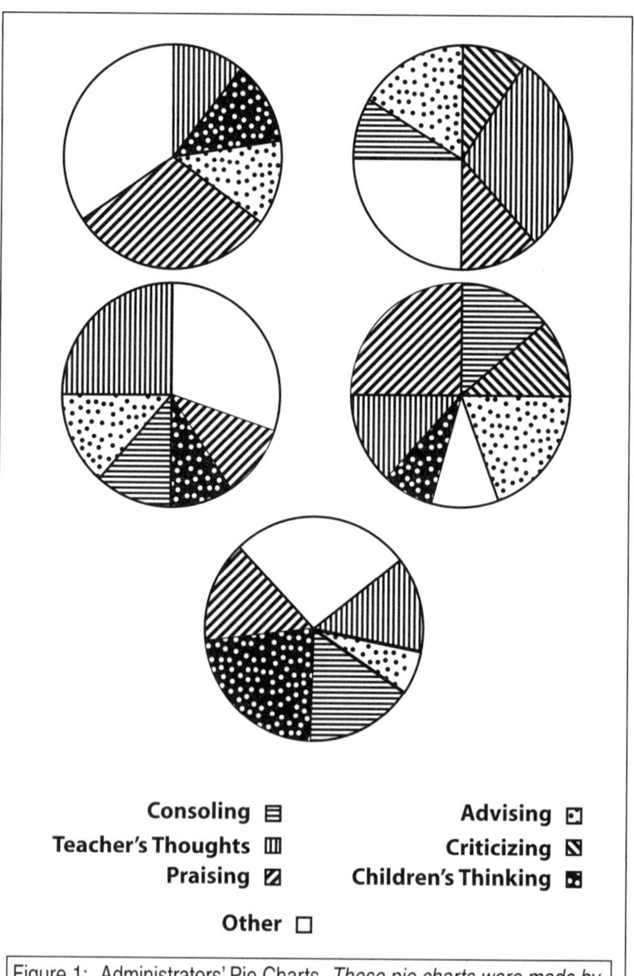

Figure 1: Administrators' Pie Charts *These pie charts were made by five different administrators, analyzing a post-observation conference.*

I didn't spend much time in sharing curiosity about children's thinking. I should have—might help probe teachers' thinking about their own practice.

That's the one spot that I personally didn't pay much attention to. If we could work to making that shift, it would help teachers make a similar shift in terms of kids' thinking. To get there, we have to do it too.

It will be interesting for you to see the range of communication styles used in your group. Keep in mind that even though the administrators in your group will have read the Franke *et al.* article for homework, it is likely that the importance of teachers' focusing on their students' mathematical thinking will not be reflected in participants' pie charts. Even if they read the article before their post-observation conference, they will most likely not yet have had the opportunity to give much thought to generative learning in their supervisory practices.

As you prepare for this activity, you might be concerned that participants could be embarrassed to be making color-coded pie charts as part of an adult professional development activity. However, in previous classes we found that administrators enjoyed the opportunity to express themselves in this way and enjoyed the camaraderie that comes out of it. (Interestingly enough, a number of participants have also appreciated the mathematical challenge of representing their data in a pie chart.)

Making Personal Charts Public

♦ What do you notice as you look at all the pie charts around the room? It may be the case that the dominant mode of communication shown in the charts is **advice-giving.** Until they see its representation on their own charts and on those of their colleagues, some participants may have little idea of how much they and their peers jump immediately to "fixing the situation" without first considering the students' thinking or the teachers' views of what the students are doing mathematically.

This may be very unsettling for participants, creating a feeling of disequilibration similar to that teachers experience when they recognize that teaching is not synonymous with telling. Participants may recognize that their own strategies and practices for conducting teacher conferences may not be the most productive or that they may even be counterproductive to the outcomes they would like to see.

As a facilitator, you will need to be aware that these reactions may occur and be ready to offer support. For example, you may wish to reassure them that their initial reactions to a particular lesson may still be on target and their advice sound and appropriate. For instance, the administrator may have found that the purpose of an observed activity was unclear or the teacher's explanation confusing. Stress that the focus is on *how* that information is conveyed to teachers. Let participants know that in a later activity they will try out different modes of communicating with teachers.

If **sharing curiosity about the students' mathematical thinking** is underrepresented on the charts, you might ask participants to consider an explanation for the little attention paid to this mode of communication.

Discussion of Individual Charts

♦ What goals did you have for this conversation? How did you make your decisions on how to approach the conversation? How did your decisions influence the conversation and the teacher's reaction? and Reflecting back on the conversation, what did you think? The set of questions about individual charts is designed to help participants think through how well their goals for the conversation aligned with what actually happened. For example, if a participant's goal was for a teacher to become curious about her students' mathematical thinking, suggesting to a teacher that she pay more attention to what her students are thinking might not stimulate the teacher's curiosity as much as engaging the teacher in thinking about what certain students might have been thinking.

Discussion of the Charts as a Whole

♦ Are there several different kinds of profiles represented here? From what circumstances did these profiles arise, and what was the result? Here administrators look at the different profiles that are represented in the charts and query the producers of them. They might ask questions such as the following:

> What happened when you spent most of the time giving advice?
>
> With which teachers did you tend to do this? How did these teachers react?
>
> What happened for participants who engaged in problem-solving with teachers?
>
> What happened for participants who shared a curiosity about student thinking with the teacher?
>
> Under what circumstances did you tend to use these approaches, and what was the result?

Toward the end of the discussion, ask participants to think about what trends they see emerging from looking at all charts as a whole.

We believe that administrators' communication patterns with teachers can benefit from less direct advice-giving in the conventional sense because that mode of communication tends not to be the most effective to help teachers construct knowledge. Still, we are not advocating one "right" pie chart. The actual configuration of communication modes will differ depending on the situation, and each administrator will need to develop the sensitivity to know what is most appropriate in a given situation. There may be times when inquiring into what the teacher thought about the students' mathematical thinking may feel too invasive to the teacher, or there may be times when giving advice is the right thing to do. Participants may raise such points themselves, and we encourage you to acknowledge and validate them.

Consolidating New Understanding

If there is time in this session, suggest that participants revise their pie charts to reflect new thinking they may have done during this activity, taking into consideration the needs of the teacher, the goals they have for the conversation, and the approaches that would best support the teacher in his or her development as a generative learner. Later in the session, participants will engage in the role-play activity to help them think specifically about supervisory practices that help teachers become generative learners.

ACTIVITY 2
Video
65 minutes

Materials
Videotaped Clip 5
Pattern blocks
Handout 27
Flip chart and markers

Focusing on the Mathematics Content of Classes

15 minutes

Discussing Classroom Observations Ask for volunteers to describe what they found when they did a classroom observation using the Math Content Observation Guide. Invite them to talk about the questions on the Guide and more generally about their experiences observing for content.

Using the questions about mathematical content, what did you find out?

Remind participants to be attentive to issues of respect for where teachers are in their understanding of mathematics and mathematical teaching. These issues are discussed in the **Introduction** of this course.

15 minutes

Doing the Math Tell participants that before they watch the next classroom episode, they will do the mathematics that the students are doing in the episode. Pass out the pattern blocks, and ask participants to find as many different ways as possible to cover a hexagon with the other shapes. After a few minutes, ask the following discussion questions:

How many ways are there to cover the hexagon?

How can you prove that you found them all?

What are the fractional relationships among the pieces?

35 minutes

Viewing and Discussing the Videotaped Classroom Episode (16:08–28:50) Tell participants that they will now watch the videotaped clip *Pattern Blocks,* in which students explore the same mathematics that they just did. Give participants the following background information about the classroom:

- This is a second grade class.
- Students in the class are conducting an initial exploration of geometry.
- The 14-minute episode has been edited to show portions of the introduction, small-group work, and whole-group discussion of the investigation. Occasionally one also hears the teacher, Ms. Christiansen, in a voice-over reflection.

Tell participants that they will watch the videotaped clip just once. Suggest that they take some notes about the questions on the Math Content Observation Guide as they do.

Before showing the videotaped clip, ask participants to think carefully about the orientation they bring when they view a classroom episode. Encourage them to resist the temptation to make broad judgments about the teacher and the classroom on the basis of the very limited view they will get. You might say:

The central purpose for viewing and discussing videotaped clips in this course is to stimulate thoughtful discussion about learning, teaching, and mathematics that are sparked by watching the videotape. It is not to critique the teaching. Teaching is complex, and from the perspective of an outside observer there are always other moves that might have been made. Yet an outside observer, particularly one who sees only a short excerpt of a lesson such as what we will be seeing, will not know enough about the background of the class, the context of the lesson, or the teacher's thinking to make valid judgments about whether the teacher's decisions were good ones.

Pass out the Math Content Observation Guide. View the video, and then discuss the following questions, which are taken from the Math Content Observation Guide and are particularly appropriate for this video.

What specific ideas are being explored by different students?

What kinds of mathematical thinking are happening (conjectures, proofs, revising, generalizing, and so on).

How does Ms. Christiansen show that she understands enough about the mathematics to test the boundaries of the students' understanding?

Is the teacher responding to the mathematics in the students' mathematical thinking? If so, how?

Let participants know that in the next activity they will turn their attention to the idea of generative learning.

Discussion of Classroom Observations

Participants discuss observations that they made using the Math Content Observation Guide. As background for these discussions, be sure to mention the issues about *Orientation Toward Observing Teachers* described in the **Introduction** (p. 10), in particular the notion of not jumping to conclusions but thinking about what the teacher understands and what he or she needs to learn. Again, the stance of making conjectures and citing evidence for those conjectures will be helpful in maintaining an attitude of inquiry.

- **Using the questions about mathematical content, what did you find out?** Given the current state of affairs, it is quite likely that participants will have observed classrooms in which the teacher's mathematics knowledge is fragile and in which the teacher does not spend much time working with students' mathematical ideas. It may be helpful to remind participants who had this experience that most teachers were taught mathematics in the traditional way, which does not promote deep understanding of mathematical ideas. Teachers themselves may feel bad about this and may be worried about their capacity to serve their students well. Such teachers would probably appreciate suggestions about how to improve their mathematics knowledge. This will be an opportunity for the participants to discuss what teachers need to know about elementary mathematics in order to teach it well and how they might learn those mathematical concepts. Some possibilities are participation in a professional development program that provides the opportunity to rethink elementary mathematics, becoming interested in the way students are doing the mathematics and exploring it further on one's own, and working with other teachers in a study group to examine students' mathematical work and understand it thoroughly.

Administrators may not have been able to respond to all of the questions on the guide, either because there was so much to keep track of or because there was no evidence for some of the questions. It is important to remind participants that no evidence does not mean that there might not be evidence in another class session. Pre- and post-observation conferences are meant to provide opportunities for administrators to learn about a teacher's practice.

Doing the Math

Participants first do the mathematics that the students in the videotaped clip are doing: find as many ways as possible to cover a hexagon with the other pattern block shapes. The discussion questions are as follows:

- **How many ways are there to cover the hexagon?** The students on the videotape found nine ways to cover the hexagon. They counted the hexagon itself and discovered that there are two ways to represent $\frac{1}{3} + \frac{1}{3} + \frac{1}{6} + \frac{1}{6}$.
- **How can you prove that you found them all?** If you have used all combinations of halves, thirds, and sixths, you have found them all.

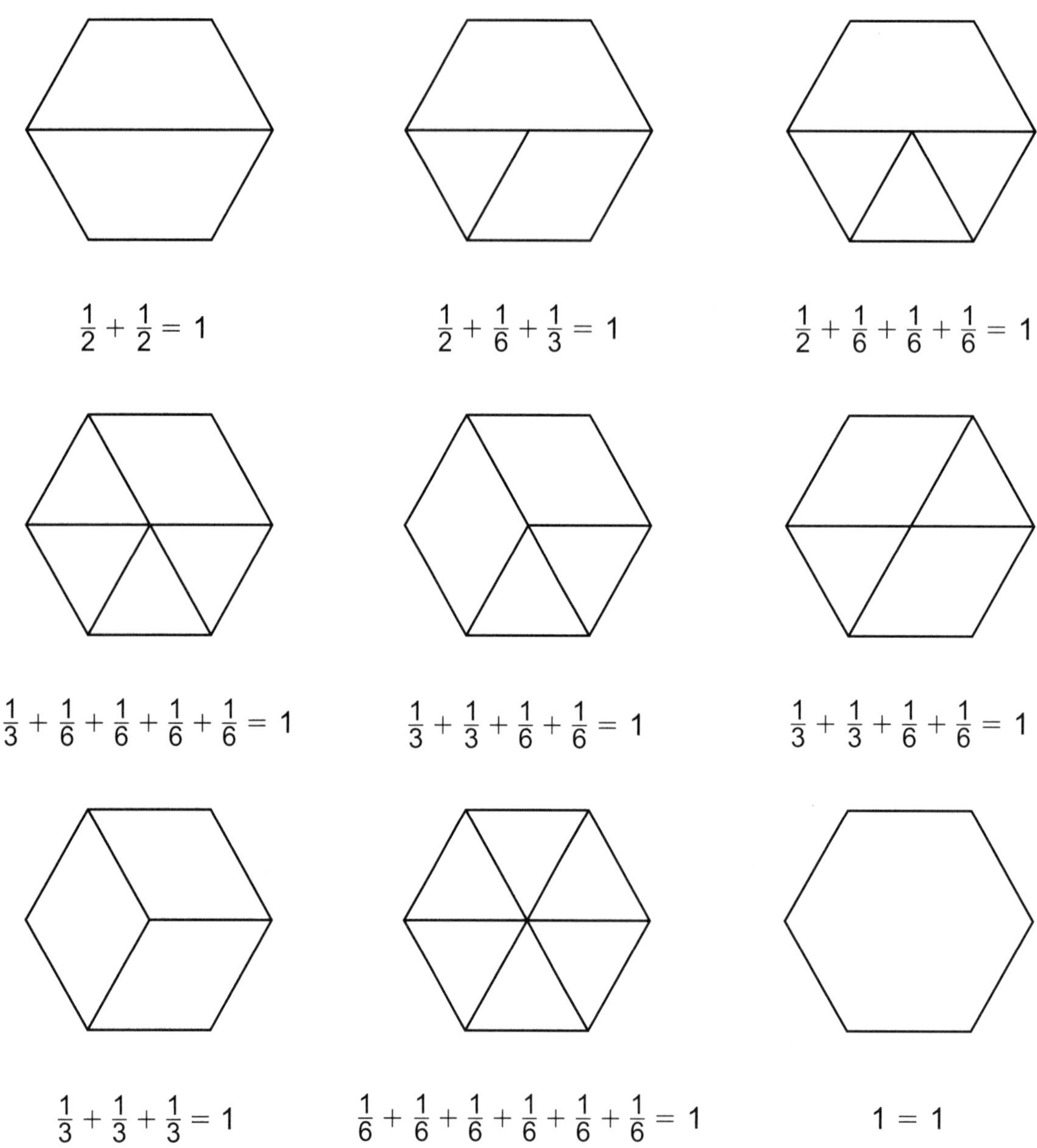

Figure 2: Pattern Block Combinations

• What are the fractional relationships among the pieces? As you will see when you play with the pattern blocks yourself, there is a systematic relationship among the pieces. If the hexagon represents 1, each trapezoid is $\frac{1}{2}$, each rhombus is $\frac{1}{3}$, and each triangle is $\frac{1}{6}$. This means that, using equivalent fractions, it is possible to work out all of the possible combinations.

Viewing the Videotaped Clip

This segment, *Pattern Blocks,* depicts the teacher, Ms. Christiansen, introducing the topic of geometry to her second grade class. This video, like the one shown in the last session, was made at the Lincoln School in Brookline, Massachusetts. Ordinarily we would not show two clips from the same school, but we wanted to show a primary class at some point in this course, and *Pattern Blocks* was the most appropriate episode available.

Ms. Christiansen encourages the students to play freely with the pattern blocks for a while, saying that because the blocks are new for the students, they would play with them anyway. She then introduces the students to the idea that each geometric shape has its own name. She asks them what relationships among the various shapes they discovered while they were playing with them, and she challenges them to play a game to find as many different ways as possible to cover a hexagon with the other pieces. This episode includes voice-overs in which Ms. Christiansen describes her intent at various points in the lesson.

Because this session is so full, participants will view this video only once. They will delve deeply into the math content of the class using the Math Content Observation Guide. We have selected a few of the bulleted items for discussion here because they are particularly appropriate to this episode.

• What specific ideas are being explored by different students? As in the last episode, *Fraction Tracks,* participants should understand that this question is not asking them to identify what students do not understand (as though there were a situation that needed to be fixed). Rather, they are being asked to acknowledge that the students in this (and any) class are working on several aspects of the mathematical ideas presented and to be quite descriptive about these ideas. In this classroom episode, we see students working on the names of the shapes (students seem to remember the names of some shapes but not others). They also are working on describing the characteristics or properties of the shapes—the number of sides and corners each has. Many students are working on the relationships among the shapes; for example, that two trapezoids equal one hexagon. One student, Ean, is working on the relationship between the shapes of the pieces and their fractional values when she notices that each rhombus is $\frac{1}{3}$ of the hexagon and each triangle is $\frac{1}{6}$ of the hexagon. Finally, several students are working on ways of knowing that they have found all of the possible ways to cover the hexagon, some of which are more mathematically valid than others. One student reverses what his partner did; another tries as many ways as he can, always gets nine, and so concludes that nine is the right answer. Christian develops a more systematic way by starting with the big pieces, then combining big and small pieces, and finally going to the small pieces. This is essentially a physical version of the process of working out all of the equivalent fractions. (Note that he says that there are nine rather than seven ways to do it because he is counting the hexagon itself and two different representations of $\frac{1}{6} + \frac{1}{6} + \frac{1}{3} + \frac{1}{3}$).

♦ **What kinds of mathematical thinking are happening (conjectures, proofs, revising, generalizing, and so on)?** The students are engaged in *exploring* the many ways the pieces can be made equivalent to a hexagon, *observing* how the pieces are related, *constructing* hexagons out of pattern blocks, and *proving* that they found all of the ways to cover the hexagon.

♦ **How does Ms. Christiansen show that she understands enough about the mathematics to test the boundaries of the students' understanding?** There are a number of examples of this. The most striking occurs during the interview at the end of the videotaped clip when Ms. Christiansen talks about the relationship between geometry and number sense, which she sees as "webbed through" geometry. She cites the student who talked about thirds and sixths and notes that when they do symmetry, it will involve doubles.

♦ **Is the teacher responding to the mathematics in the students' thinking? If so, how?** Ms. Christiansen is very responsive to the students' mathematical thinking. Throughout the lesson she asks them (individually and collectively) questions that are not generic but depend on knowing the mathematics and having listened to the mathematics in the students' thinking.

Can you tell us how many sides it has?

Can you feel how many corners it might have?

Who can draw a picture of that piece?

Do you know the name of the piece?

ACTIVITY 3
Discussion of Reading
20 minutes

Materials
Handout 22
flip chart paper
markers

What Is Generative Learning?

20 minutes
Whole group

Discussing the Franke *et al.* Article Participants read *Capturing Teachers' Generative Change: A Follow-up Study of Professional Development in Mathematics* by Megan Loef Franke, Thomas P. Carpenter, Linda Levi, and Elizabeth Fennema and answered a set of questions for homework.

Let participants know that this reading about generativity will lay the groundwork for rethinking the work they do with teachers that began in this session and will continue in subsequent sessions.

Begin by inviting participants to react in a general way to this research paper. You might ask:

What struck you as you read this research report?

After a number of participants have responded, bring out some of the main ideas by focusing the discussion on each of the homework questions in turn.

How do the authors characterize generative learning?

What view of teachers-as-learners do the authors articulate?

What do teachers need to know to be generative learners?

Why does a focus on children's mathematical thinking help promote generative learning in teachers?

Keep track of participants' ideas on flip chart paper, and help flesh them out where necessary. These ideas may be useful to refer back to in subsequent sessions.

Introduction

This article was written for a research audience rather than a practitioner audience, and some participants may find it more technical than the articles they usually read. It was chosen for the ideas about teacher learning it describes and the the discussions it is likely to generate. The ideas presented can help participants think through how their supervisory practices can support teachers' ongoing learning about their students' mathematical thinking.

It is important for administrators to realize that teachers do not need to be generative learners to be working in effective ways on their classroom practice. This is a very advanced stage of teacher development. Even after participating in the *Cognitively Guided Instruction* (CGI) professional development and reflecting on its principles for a number of years, fewer than half of the teachers in the Franke *et al.* study were considered generative learners. Thus, it is not a goal of this session for participants to think that the teachers in their schools *should* be generative learners. Rather, the idea of generative learning can serve to help participants reframe the work they are doing with teachers and can set both teachers and administrators on a path to becoming generative learners.

Summary of the Research Report

This research report describes a study the authors undertook to investigate the extent to which a group of 22 teachers who participated in the same professional development program continued to implement the principles of the program four years after its completion. The goal of the *Cognitively Guided Instruction* (CGI) program is for teachers to use their classrooms as contexts for developing their understandings of students' thinking. Teachers are given research-based knowledge of how students' mathematical knowledge develops and observe that students in their own classrooms solve mathematical problems in similar ways.

A central tenet of the program is that teachers learn about students' mathematical thinking by listening closely to their ideas and working to understand them. From such insights, teachers consider what pedagogical steps to take next while they continue to explore students' thinking. Teachers not only learn about individual students' mathematical thinking in this way, but they also come to observe general patterns in how students' mathematical thinking develops confirming research they have read.

Some teachers come to realize that they can make the research knowledge their own, and even extend it, by continuing to think about their students' ideas, figuring out what their students' responses reveal about their understanding. When teachers engage in a process of ongoing learning such as this one, they are engaging in *generative learning* (also referred to as "generative growth" or "generative change" in the paper). According to the authors, teachers who engaged in generative learning "(a) viewed children's thinking as central, (b) possessed detailed knowledge about children's thinking, (c) perceived themselves as creating and elaborating their own knowledge about children's thinking, and (d) sought colleagues who also possess knowledge about children's thinking for support" (p. 77).

Franke *et al.* developed a continuum of levels of teachers' engagement with students' mathematical thinking four years after the CGI program ended. On the basis of the criteria that they attributed to each level, they assigned each of the participating teachers to a level. The authors report that the practice of all of the teachers (22) involved in the program continued to reflect what they had learned in the program to some extent. Nine of the teachers who ended the CGI program with a high degree of engagement with student thinking sustained this high level four years later. These were the teachers who were engaging in generative learning. (See pages 85–86 and p. 88 in the book of *Readings* for a comprehensive description of teachers' levels at the end of the program compared with their levels four years later.)

The authors also discuss what distinguishes the different levels from one another and provide examples from the data they collected about the teachers to illustrate their findings.

Discussion

♦ What struck you as you read this research report? This first question gives participants a chance to react in a general way to the article and provides an opportunity for you to learn what participants find interesting and/or provocative about the study.

The other discussion topics are the questions participants answered for homework. Because the ideas in this article lay the groundwork for the thinking that participants will explore more fully throughout the rest of the course, the goal of this whole group discussion is to get the ideas out. Later in the session, participants will consider the implications of these ideas for their practice as supervisors.

♦ How do the authors characterize generative learning? Franke *et al.* refer to generative learning as the ability of individuals to continually add to what they understand. It is an iterative process in which one's knowledge base grows and is refined as one makes use of this knowledge in new situations. The process more explicitly consists of applying knowledge to a new situation, reflecting on what transpires, planning next steps on the basis of new information, and observing and making sense of the result.

♦ What view of teachers-as-learners do the authors articulate? Franke *et al.* make a distinction between the kind of professional development that can help teachers develop as generative learners and that in which "fidelity of teachers' practices to those specified in the training program" (p. 81) is the goal. This relates to Nolan and Francis' view of teachers as constructors of their own knowledge rather than as enactors of particular teaching practices. In order to engage in generative learning, teachers need to be researchers and problem solvers as they make sense of their students' mathematical ideas and cultivate the judgments that support them in making productive instructional choices in their day-to-day practice in the classroom.

The kind of supervisory practices that would best support generative learning is a topic that participants will address later in the session.

♦ What do teachers need to know and be able to do to be generative learners? Teachers need to bring to their work a curiosity about their students' thinking and a desire to make sense of it. They need to appreciate the opportunities for ongoing learning they have in their classrooms and be able to take advantage of them by probing and making sense of their students' thinking, figuring out what the best next pedagogical move would be, and taking stock of the decision they made on the basis of their students' responses. Teachers need to bring a growing understanding of mathematics and children's mathematical thinking plus the commitment to further their understandings to their work in the classroom.

♦ Why does a focus on children's mathematical thinking help promote generative learning in teachers? When teachers probe their students' mathematical thinking, they are making that thinking available for their own learning. They can observe trends in their students' thinking and look for structures to increase their understanding of the development of students' mathematical ideas. They can try a particular strategy as they work to push the boundaries of their students' thinking and determine its effectiveness as they listen to how their students respond. Franke *et al.* also point out that a focus on children's mathematical thinking allows teachers to talk to colleagues and other staff about their students' ideas, and in this way support the generative learning of school communities where everyone, students, teachers, and administrators, are learners.

ACTIVITY 4
Role Play and Discussion
40 minutes

> ***Materials***
> Handouts 25 & 26

Taking Different Orientations Toward Talking with Teachers

5 minutes

Setting up the activity Tell participants that they will now have a chance to adopt different orientations to communication in a role play of a post-observation conference with a teacher. Mention that the videotaped clip they just viewed—Ms. Christiansen's second grade class—is the one that they will be discussing.

Ask participants to pair up and decide who will take on the role of supervisor and of teacher. Tell them that you will give the supervisor a scripted set of questions to incorporate into the role play. The supervisor can weave the questions together with improvised text, and the teacher can improvise responses as seem appropriate.

> In this activity, participants experiment with several different ways of talking with teachers after a classroom observation. Doing a role play allows them to experience what it feels like to be the supervisor and the teacher in different scenarios.

15 minutes

Doing the Role Plays Pass out Handout 25, Taking Different Orientations, which contains the scripts for the role plays.

As pairs complete their role plays, pass out Handout 26, and ask pairs to reflect on their experiences during the role play, jotting down notes to share later with the whole group about the following questions:

Questions for the Teacher

What do you gain from this approach to communication?

What do you lose?

Questions for the Supervisor

What do you gain from this approach to communication?

What do you lose?

If there is time, invite participants to switch roles and do the series of role plays again. If they would like, supervisors can develop their own comments and questions for the post-observation interview. Again, ask them to reflect on their experiences on Handout 26.

20 minutes

Whole-group discussion

Discussing the Role Plays Call participants back together for a discussion of their experiences during the role-plays. Invite them to reflect on each mode of communication, from both the teacher's and the supervisor's perspective. Ask such questions as the following:

> *What were benefits and trade-offs of the advice-giving mode? Of sharing curiosity about students' thinking? Of asking questions and finding out what the teacher intended? Are there situations in which one approach might be more beneficial than another?*
>
> *What does each of these approaches require of the supervisor in terms of prior knowledge?*
>
> *What does a broader range of approaches to communicating with teachers require that the supervisor attend to and do during the observation itself?*

Leave some time at the conclusion of this session to pose the following question in order to encourage participants to think about collaborative inquiry in the context of generative learning.

> *How does sharing curiosity about students' mathematical thinking align with supervisory practices that support generative learning?*

As participants discuss the role plays, listen for what they might be saying about the opportunities to learn about mathematics, teaching, learning, and students' thinking that engaging in collaborative inquiry with teachers can provide. In Session 8, participants will consider the idea of the school as a learning community for teachers as well as students.

Setting up the Activity

This activity gives participants the opportunity to experiment with several different ways of talking with teachers after a classroom observation. Doing a role play allows them to experience what it feels like to be the supervisor and the teacher in several different scenarios.

Doing the Role Plays

Each of these three interview scenarios has been constructed to be a plausible response to the lesson on the videotape. The *advice-giving* scenario picks up on several aspects of Ms. Christiansen's open-ended style of teaching and explores the consequences of such a style. The *sharing curiosity about children's thinking* scenario focuses on the supervisor (or classroom observer) and teacher exploring together the mathematical ideas at play in the classroom. The *finding out what the teacher intended* scenario holds back on making any recommendations in order to explore with Ms. Christiansen the pedagogical moves she made during this class.

Having the opportunity to try different orientations and get a sense of the different dialogues one can get from each can be eye-opening. Point out to participants that each orientation will also be affected by elements from the cultural backgrounds that teacher and supervisor carry with them, such as the tone and volume of their voices and body language.

Engaging in these role plays may also bring home to participants the critical role of listening for the intellectual ideas at play in inquiry-based observations. Although the three scenarios are all based on plausible interpretations of Ms. Christiansen's lesson, the second and third scenarios rely on having grasped some of the fundamental mathematical issues with which the students are grappling.

Another point the scenarios make is that the professional development relationship between teacher and supervisor has parallels to the relationships that teachers develop with the students in their classrooms. As part of the discussion of this activity, you may wish to raise this point if others do not.

You may need to be prepared for the possibility that some participants might find the collaborative inquiry modeled in the second and third scenarios challenging. They might not fully understand the mathematical questions that are posed and could find it difficult to role-play. You can invite people to ask for clarification if they do not understand a point or a question.

♦ **What were the benefits and trade-offs of the advice-giving mode? Of sharing curiosity about children's thinking? Asking questions and exploring what the teacher intended? Are there situations in which one approach might be more beneficial than the others?** The following points are relevant to this discussion:

- When *advice giving* is the dominant mode of communication, teachers are not supported in constructing their own knowledge and supervisors cut themselves off from being learners themselves. Advice giving may be the most appropriate way to transfer information when a supervisor has experience and knowledge of something that it would be important for a teacher to be aware of.
- In contrast, when *sharing curiosity* is the dominant mode, teachers and supervisors think together about events related to the mathematics from the class session that really puzzled them and push each other's thinking forward. If the puzzlement is not genuine for either the teacher or the supervisor, then the shared aspect of the experience may be lost.
- The approach of *finding out what the teacher intended* can help participants make better sense of the class they are about to see (if in the context of a pre-observation conference) or already saw

(in the context of a post-observation conference). Information about the teacher's intent can help administrators suspend their judgments about what took place. For teachers, questions about intent can help clarify what it is that they hoped to accomplish. On the other hand, questions about intent can make the person being questioned feel defensive about the choices he or she made.

You may wish to note that, although each of these scenarios has been constructed to illustrate a point, tone is also an important feature. *Advice giving* could come off sounding like a command to do something, or it could sound like a good-faith effort to help out. On the other hand, *asking questions and finding out what the teacher intended* could end up sounding like an interrogation and make the interviewee feel judged (e.g., "Why are you asking me that?"). In fact, teachers who are learning to ask students about their thinking have said that students can react defensively to such well-intentioned questions. Thus, it is important for administrators to monitor the tone and style they use as they role-play each of these scenarios.

As participants consider these three scenarios for their actual practice of teacher supervision, it will be important to acknowledge that it can take time to cultivate the skill of asking probing questions. You may also want to reemphasize that the discernment needed to figure out the questions to ask is cultivated from the kind of careful listening that has been part of all the sessions of this course.

In order to help participants think about making a shift in how they interact with teachers, you might mention that one place to start is to ask genuine questions—that is, questions to which they really do not know the answer. For example, if they really did not have a sense of what the teacher intended by doing a particular lesson or if they really did not understand a student's explanation, then they could ask about it: "I was really confused by what that student was trying to say. What do you think her point was?" "I'm not that familiar with the mathematics in this lesson. Why do students work on it in this grade?" This kind of genuine questioning on the part of the observer can encourage teachers to be in touch with their own curiosity about students' mathematical thinking and pursue it in their teaching. As was noted earlier, such curiosity is an essential feature of generative learning.

♦ **What does each of these approaches require of the supervisor in terms of prior knowledge?** Although all of these approaches would benefit from supervisors having knowledge of the mathematics that students are working to understand, for some the focus was more on issues that have little to do with the mathematical content of the class. In the mode of *advice giving,* for example, the questions about the choices that the teacher made did not relate to the mathematical content of the class as much as the questions that were posed in the *sharing curiosity about children's mathematical thinking* mode.

♦ **What does a broader range of approaches to communicating with teachers require that the supervisor attend to and do during the observation itself?** Because the idea of *sharing curiosity about children's mathematical thinking* is probably new to most participants, they may not be in the habit of paying close attention to how students are thinking and how the teacher is working with students' ideas. Thus, to support teachers who are trying to help students develop their mathematical thinking, participants will need to pay attention to what students are saying and formulate questions that would help the teacher inquire into students' thinking in enough depth to be able to work with their ideas.

♦ **How does sharing curiosity about students' mathematical thinking align with supervisory practices that support generative learning?** As has been discussed above, both *advice giving* and *asking about the teacher's intent* can be appropriate and useful modes of communicating with teachers. However, there are several points to make about what can be gained when using collaborative inquiry modes of communication that other modes can not provide.

- **It supports teachers as they construct their own knowledge.** *Advice giving* is certainly an important and necessary mode of communication in some situations, but it may not be the most effective because it does not encourage or support teachers' construction of their own knowledge. *Sharing curiosity about children's mathematical thinking,* on the other hand, does support teachers' construction of their own knowledge. Participants discussed this idea earlier in the session, and it may be worth revisiting it here.
- **It helps teachers better understand their students' thinking and allows administrators to take part in important conversations about mathematics instruction in their schools.** Supervisors who use communication modes that are more collaborative in nature can support teachers to focus on students' mathematical thinking and to reflect on how to work with students' ideas. Developing an understanding of the mathematics that teachers and students are working on together allows administrators to be part of important conversations about the learning and teaching of mathematics in their schools and districts. Furthermore, supervisors whose stance is one of collaboration are in a good position to increase their *own* knowledge of mathematics, teaching, and learning.

Another important point to make in this discussion relates to the effect of generative learning on students. Teachers who are engaged in generative learning are working to bring an understanding of mathematics and of students' mathematical thinking to their work with students. These understandings contribute to the ongoing development of their students' ideas. When administrators support teachers through their supervisory practices, there is learning about mathematics and students' thinking happening at all levels, for students, teachers, and administrators themselves.

As participants discuss the experiences they had doing the role plays, listen for what they might be saying about learning about mathematics, teaching, learning, and students' thinking that engaging in collaborative inquiry with teachers can provide. In Session 8, participants will consider the idea of the school as a learning community for teachers as well as students.

CLOSING
15 minutes

Materials
Handout 28

Homework

5 minutes

Assigning Homework As you assign homework, you might mention that the class has just completed the second of two sessions on observing for the mathematics content of a classroom. The next two sessions will focus on observing for the learning and pedagogy of a mathematics classroom and will continue consideration of how supervisors might talk with teachers.

Bridging to Practice

10 minutes

Reflective Writing Remind participants that Bridging to Practice is designed to allow them to focus on the ideas in the course that seem particularly interesting for them and for their school community. Point out that it provides a framework for planning what they can do in order to move themselves and their schools along with these ideas.

Post the following questions, and invite participants to respond to them:

> *Choose an idea that came up today that you found particularly interesting. What is your current thinking about this idea?*
>
> *Where is your school now with regard to this idea?*
>
> *What are one or two things that you, as instructional leader, will pursue to move yourself and/or your school along with this idea?*

Your Own Journal Writing

Within one or two days of teaching this class, after you have had the chance to unwind from the class and perhaps talk with a colleague or friend about how it went, set aside some time to write your own journal entry about the class. We encourage you to write about the ideas held by the participants in your group rather than your own actions and sense of how things worked out. In our experience, writing about what the participants are thinking will be much more useful to you as you prepare for the next session. Some facilitators use the time when administrators are writing in their own journals to make preliminary notes for this journal writing.

Taking Different Orientations

Dear Supervisor,

In this post-observation conversation, you will have the opportunity to take on three different orientations as you communicate with the teacher whose geometry lesson you just observed: 1) giving advice; 2) sharing curiosity about children's thinking; and 3) finding out what the teacher intended.

Conduct three different post-observation conversations with this teacher, using a different approach for each interview. Be sure to incorporate the statements or questions listed under each orientation into the relevant conversation.

Giving Advice

- At the beginning of the class, you gave several students slates so that they could draw pictures of the shape that the student with his hand in the bag was describing. I would suggest that next time you give slates to all of the students. That way, every student would be involved in the lesson and would have the opportunity to translate the description in words to an image on the slate.
- Although the discussion that the students had about the ways to cover the hexagon with the other pieces was very good, it might have been even better if you had pointed out that they had two ways of representing two greens and two blues. Then you could have pointed out that they were using the same pieces but arranging them in different ways, which would have been the beginning of understanding equivalent fractions.
- I think it may have been confusing that you put the ways of covering the hexagon with other pieces on a slate in front of you. I don't think all of the students could see. I would suggest that next time you use an overhead or draw the pieces on flip chart paper so that all of the students can see all of the ways of covering the hexagon.
- When Christian explained his way of making sure that he had all the ways of covering the hexagon, his explanation was very unclear. Next time, you might want to ask a student to explain a bit more so that all of the students can understand what he has done.

Sharing curiosity about children's mathematical thinking

- I was fascinated by Ean's talk about fractions. If you remember, early in the class she said that three blue ones (rhombuses) were $\frac{1}{3}$ each. She said that $\frac{1}{3}$ meant that there were three equal parts. Later she said that each of the triangles was $\frac{1}{6}$. I wonder what she really understands about fractions. What do you think she understands? Have you heard her talk about fractions before? Do you think she knows that fractions can be different sizes depending on the size of the whole? What could you do in class sometime to explore what Ean understands about fractions?
- Christian had a very interesting way of figuring out how many ways there were to cover the hexagon. From what you said in class, you seemed to think that his way was more systematic than those of the other students. He started with the big pieces, like the hexagon or the trapezoid, and then he added the small ones, like the triangles. I wonder what Christian was understanding about this particular strategy that led him to do it that way? What was systematic about what he was doing?

Asking questions and finding out what the teacher intended

- I'd like to get your thinking on how you decided to have the students reach into the bag and describe the shape they were feeling. I'm curious about what you were trying to achieve.
- I am curious as to why you let the students do this activity in such an open-ended way. You could have shown them how to do it, but you didn't. It took a long time for them to do it the way you designed it. I'm interested in getting your thinking on why you chose to do it that way.
- When you were asking the students how they knew that they had found all of the ways to cover the hexagon with the other pieces, you called on Christian, commenting that you had heard him saying something interesting. I'm curious to know what it was that you heard Christian say that made you want him to say it out loud. How did that fit with what you wanted the rest of the students to be thinking about?

Role Play

Questions for "The Teacher"

1. What do you gain from this approach to communication?

2. What do you lose?

Questions for "The Supervisor"

1. What do you gain from this approach to communication?

2. What do you lose?

Math Content Observation Guide

Students		
Focus Question	Conjectures	Evidence from Classroom
What mathematical ideas are embedded in the lesson?		
• What is the topic?		
• What are the ideas within this topic that are being explored?		
• What specific ideas are being explored by different students or groups of students?		
What makes this worthwhile mathematics?		
• What important mathematical ideas are involved?		
• What is the relationship between doing procedures and exploring ideas in this mathematics?		
• What kinds of mathematical thinking are taking place (conjectures, proofs, revising, generalizing, etc)?		

Math Content Observation Guide

Teachers		
Evidence from Classroom	Conjectures	Focus Question
		What does the teacher seem to understand about the mathematics?
		• What aspects of the mathematics does the teacher seem to know well and in what areas does he or she still seem to need to deepen her knowledge?
		• Is the teacher responding to the mathematics in students' mathematical thinking?
		• How does the teacher show that he or she understands enough about the mathematics to test the boundaries of students' understanding?
		What does the teacher seem to understand about the development of children's ideas in this topic?
		• How does the teacher show that he or she understands students' thinking?
		• What follow-up questions does the teacher ask to probe the robustness of students' understanding?
		• In what ways does the teacher help them extend or deepen their thinking?
		What seems to be the teacher's long-term mathematical agenda?
		• What mathematical ideas is the teacher probing?
		• What does the teacher do to bring into focus the long-term importance of these mathematical ideas?

Homework for Session 6

A. "Canceling" Zeroes: In the next videotaped clip, you will see a fifth grade classroom exploring the question of "canceling" zeroes when looking for an equivalent fraction. For example, is it all right to "cancel" the zeroes in $\frac{10}{30}$ to get the equivalent fraction, $\frac{1}{3}$? To help you make sense of the mathematics you will be seeing in this video, do the following mathematics assignment.

Below is a set of fractions, all of which have zeroes in the numerators and denominators. Sort these fractions into at least three categories. In one category, put the fractions for which you do not get equivalent fractions when you cross out zeroes. In the other categories, put fractions for which you do get equivalent fractions after crossing out the zeroes. For these other categories, describe what mathematical principle allows you to cross out the zeroes. Think of some rules to determine when "canceling" zeroes works.

$\frac{10}{30}$	$\frac{2001}{6003}$
$\frac{10}{101}$	$\frac{204}{408}$
$\frac{101}{201}$	$\frac{202}{205}$
$\frac{2010}{5010}$	$\frac{6200}{8700}$

B. Analyzing Discourse: In the next session, when you view the video of a fifth grade classroom, you will be focusing on the mathematical discourse that takes place among students and between the teacher and his students. To prepare you to attend to these things, please reread *What's All this Talk about "Discourse"?* by Deborah Ball (Reading 3), and find some examples of how Ball works with students' ideas. Be ready to share them during the next session.

SESSION 6

Observing How Knowledge Is Constructed in Mathematics Classrooms

> The NCTM *Professional Teaching Standards* calls unprecedented attention to the "discourse" of mathematics classrooms, as embodied in three standards: Teacher's Role in Discourse, Students' Role in Discourse, and Tools for Enhancing Discourse. An unfamiliar term to many, discourse is used to highlight the ways in which knowledge is constructed and exchanged in classrooms. Who talks? About what? In what ways? . . . Whose ideas and ways of knowing are accepted and whose are not? What makes an answer right or an idea true? What kinds of evidence are encouraged or accepted?*

In classrooms in which students are exchanging mathematical ideas, teachers are faced with the challenge of determining what to do with students' ideas after they have been aired. For teachers, from the neophyte to the experienced, these challenges are significant, and administrators who appreciate the demanding nature of what these teachers are trying to do can help teachers learn how to attend to students' mathematical ideas and use them to make pedagogical decisions.

In this session, participants shift their attention to the sense students are making of the math content and the way the teacher works with students' mathematical ideas. They are introduced to the Learning & Pedagogy Observation Guide and use it to frame their classroom observations of the teacher's pedagogical moves *in relation to the students' mathematical ideas* rather than the more typical focus on the students or the teacher alone. The interplay between students' ideas and the teacher's pedagogical moves is at the heart of classrooms in which generative learning is taking place.

Participants continue to explore ways in which the content and modes of communication of their post-observation conferences with teachers can contribute to teachers' ongoing generative learning. They consider the role of inquiring collaboratively with teachers about the mathematical thinking that takes place in their classrooms. Such curiosity about students' thinking can fuel productive discussions about how to probe students' mathematical ideas and advance teachers' generative learning.

*from Ball, D. (1991). What's All This Talk About "Discourse"? *Arithmetic Teacher,* 44–48.

Overview for Session 6

OPENING page 204 5 minutes	**WELCOME AND INTRODUCTION** This is an opportunity to let administrators know what they can expect from this session and how it connects to both previous and upcoming sessions.
ACTIVITY 1 Homework Discussion page 206 20 minutes	**EXPLORING THE MATHEMATICS OF "CANCELING" ZEROES** Participants discuss the mathematics of "canceling" zeroes that they did for homework. This activity will prepare them for viewing the videotaped clip for this session.
ACTIVITY 2 Video page 209 45 minutes	**OBSERVING THE "CANCELING" ZEROES VIDEO** Participants observe a fifth grade classroom's investigation of "canceling" zeroes two times. For the first viewing, they use the Overall Observation Guide, and for the second, they use the Learning & Pedagogy Observation Guide to focus on the interactions between students' mathematical ideas and how the teacher works with them.
ACTIVITY 3 Small group Analysis page 216 45 minutes	**MAKING SENSE OF THE LEARNING AND PEDAGOGY IN MR. PUNZAK'S CLASSROOM** Participants work in small groups to analyze a section of the *"Canceling" Zeroes* transcript, paying particular attention to the pedagogical moves made by the teacher in relation to students' mathematical ideas. Using the Learning & Pedagogy Observation Guide, they consider what questions they might like to ask the teacher in the context of a post-observation conference.
ACTIVITY 4 Whole group Discussion page 224 45 minutes	**AN IMAGE OF COLLABORATIVE INQUIRY** Participants continue to build their skills in collaborative inquiry as they develop questions to probe students' thinking. They view a videotaped conference between Mr. Punzak and a mathematics supervisor whose supervisory practices are designed to support his generative learning. The tape provides an example of collaborative inquiry about the mathematics understandings articulated in this class.
CLOSING page 229 20 minutes	**HOMEWORK** Participants observe two mathematics lessons in their schools or districts. For each observation, they will have pre- and post-observation conferences with the teacher. They use the Learning & Pedagogy Observation Guide during the observations and try out a collaborative inquiry approach in talking with the teachers during the post-observation conferences. **BRIDGING TO PRACTICE** Participants finish the session with journal writing that allows them to develop further some of their thinking from the session.

Big Ideas

Participants explore:

- students' mathematical ideas as a central focus of a mathematics class
- the teacher's central role in shaping the discourse in mathematics class
- collaborative inquiry as a way to support teachers' generative learning

---

Materials

- ☐ nametags
- ☐ flip chart paper or overheads
- ☐ markers
- ☐ agenda
- ☐ calculators
- ☐ Handout 28, Homework for Session 6
- ☐ Handout 29, Working with the "*Canceling" Zeroes* Transcript
- ☐ Reading 3, *What's All This Talk About "Discourse"?* by D. Ball
- ☐ Handout 30, Reflecting on Co-Inquiry
- ☐ Handout 31, Overall Observation Guide
- ☐ Handout 32, Learning & Pedagogy Observation Guide
- ☐ Videotaped Clip 6, *Is It O.K. to "Cancel" Those Zeroes?* (DVD #3)
- ☐ Reading 8, "*Canceling" Zeroes* Transcript
- ☐ Videotaped Clip 7, *A Conference with Mr. Punzak* (DVD #3)
- ☐ Handout 33, Homework for Session 7
- ☐ Handout 34, Pre-observation Conference Questions

Preparation

- Prepare an agenda for the session.
- Prepare the prompts for Bridging to Practice.
- Think about which discussion questions would most benefit participants, and use these to guide your discussions.
- Do the mathematics activity that was assigned for homework for Activity 1.
- Preview Videotaped Clip 6 for Activity 2. Review the notes on setting a climate for viewing and discussing videotapes in the **Introduction** (p. 9). Plan to leave enough time for participants to view the video two times.
- Prepare a large version of the Overall Observation Guide on a sheet of chart paper to collect participants' observations after the first viewing of the classroom episode.
- Work through both sections of the transcript with the Learning & Pedagogy Observation Guide in hand for Activity 3. Draw two columns on a sheet of chart paper with headings *Students* and *Teacher* to use for the discussion of the transcript. Reread *What's All This Talk About "Discourse"?* by Deborah Ball (Reading 3).
- Preview Videotaped Clip 7 for Activity 4, and prepare some questions that might be fruitful in a collaborative inquiry about students' mathematical thinking.

OPENING
5 minutes

Materials
Name tags
Agendas

Introduction

Before starting the session, post the agenda in a place where everyone can see it.

5 minutes

Introducing the Session Review the agenda briefly with participants. Let them know that in today's session they will shift their focus from the *content* of the mathematics lesson and the teacher's understanding of that content to the *mathematical sense* students are making of the content and the ways the teacher works with students' ideas. Tell participants that they will be working with the Learning & Pedagogy Observation Guide and becoming familiar with what to look for when attending to students' mathematical sense making and with how this sense-making is facilitated by the teacher.

Overview

Participants will likely feel like they are really getting the idea of how to observe differently in standards-based classrooms. You may notice that participants are now engaging with ideas in a deeper and more sustained way. As facilitator, you should be watching for this shift. Try to take advantage of it to push participants' thinking about their roles as instructional leaders even further.

In this session, the focus shifts from looking at the mathematical content to looking at the nature of the learning and pedagogy in the classroom. The **Facilitator Notes** help make the connection between the mathematical ideas of the lesson, the ways the students grapple with a complex idea, and the pedagogical moves a teacher chooses to make. Your job as facilitator is to help participants understand how interconnected content and pedagogy are. Remember again to use the Notes to prepare yourself for the possible ways the discussions can go. Think of the discussion questions as starting-off points, and be primed to move in a number of possible directions, depending on how the discussions unfold.

ACTIVITY 1
Homework Discussion
20 minutes

> *Materials*
> Chart paper and markers
> Calculators
> Handout 28

Exploring the Mathematics of "Canceling" Zeroes

20 minutes

Discussion of Homework Tell participants that they will discuss the math they did for homework. Ask them to contribute to several lists: one for those fractions for which "canceling" zeroes does not produce equivalent fractions, and at least two for which "canceling" zeroes does produce equivalent fractions. Each of these latter lists should contain fractions that follow the same mathematical rule. Record participants' responses on chart paper.

> There are two important goals of this activity: participants become comfortable with the mathematics so that they can easily follow what is happening in the videotaped episode, and they experience the intellectual nature of a mathematics classroom in which mathematical ideas are being shared.

Turn to the lists of fractions for which you get equivalent fractions and ask:

What is happening mathematically when "canceling" the zeroes here?

How can you prove that the fraction you get after the zeroes are "canceled" is equivalent to the original one?

Remember that the fractions $\frac{204}{408}$ and $\frac{2001}{6003}$ are special cases. "Canceling" the middle zeroes is not connected to division by a power of 10 the way it is for fractions whose zeroes are at the end of the numerator and denominator. Rather these fractions share a different common factor. Encourage a range of participants to share their ideas about the underlying mathematics. Ask whether there are any other situations in which you get equivalent fractions after you have "canceled" the zeroes. Explore what is happening mathematically in these cases.

Turn to the fractions for which you do *not* get equivalent fractions and ask:

What is the mathematics behind those fractions for which "canceling" does not work?

Finally, ask participants about the rules they thought of:

What are some rules you came up with about "canceling"?

Which of these rules apply to all of the fractions on the homework sheet?

Will these rules apply to all fractions with zeroes in the numerator and denominator?

Record participants' rules on chart paper, and star those that participants think apply to all of the fractions on the homework sheet. This would be a good time to point out the importance of students developing the ability to generate rules or make generalizations from specific cases in mathematics class.

Overview of the Mathematics of "Canceling" Zeroes

"Canceling" zeroes is a short-cut method for finding an equivalent fraction by dividing both numerator and denominator by the same power of 10. When both the numerator and denominator are multiples of 10, 10 is a factor of both the numerator and denominator. As such, it can be "canceled." For example, in the case of:

Does $\frac{10}{30}$ equal $\frac{1}{3}$?

$$\frac{10}{30} = \frac{1 \times 10}{3 \times 10} = \frac{1}{3} \times \frac{10}{10} = \frac{1}{3}$$

Even though zero is not the factor being "canceled"—ten is—the terminal zero is taken to represent a factor of ten, and "canceling" this 10 from both numerator and denominator is referred to as "canceling" zeroes. In this case, because 10 is a factor of both 10 and 30, the zeroes can be "canceled," resulting in $\frac{1}{3}$.

"Canceling" is sometimes referred to as *"reducing"* as it is in the videotaped clip and in many standard textbooks—as a synonym for "finding lowest terms." But *"reducing"* can be a misleading term in that it suggests the idea of decreasing the value of the fraction. In fact, when "canceling" zeroes, the new fraction is equivalent in value to the original one. The numerals in both the numerator and the denominator are smaller in value than in the original fraction, but the ratio remains the same. Thus, finding an equivalent fraction in simpler terms is a more accurate description of the mathematics underlying what is happening when "canceling" zeroes.

Discussing the Math Homework

In the mathematics activity done for homework, participants worked with eight different fractions, each of which had at least one zero in its numerator and one in its denominator. They were asked to think about for which fractions "canceling" zeroes produces equivalent fractions and for which it does not and to explain the mathematics for each category. Participants were also asked to come up with rules that govern when it is O.K. to "cancel" zeroes.

The fractions participants worked with are $\frac{10}{30}, \frac{10}{101}, \frac{10}{201}, \frac{2010}{5010}, \frac{2001}{6003}, \frac{204}{408}, \frac{202}{205}$, and $\frac{6200}{8700}$.

What is happening mathematically when "canceling" the zeroes here? How can you prove that the fraction you get after the zeroes are "canceled" is equivalent to the original one?

Participants will most likely note that for the fractions $\frac{10}{30}, \frac{2010}{5010}$, and $\frac{6200}{8700}$, both the numerators and denominators are divisible by 10 or 100.

- $\frac{10}{30} = \frac{1 \times 10}{3 \times 10} = \frac{1}{3}$
- $\frac{2010}{5010} = \frac{201 \times 10}{501 \times 10} = \frac{201}{501}$
- $\frac{6200}{8700} = \frac{62 \times 100}{87 \times 100} = \frac{62}{87}$

Because the numerators and denominators were divided by the same number (thus by 1, the identity element), the values of the resulting fractions remain the same. Participants may also mention that the numerators and denominators have factors in common.

Another way to prove that two fractions have the same value is to compare their decimal equivalents. However, this strategy can be impractical to use in cases in which the decimal equivalents yield long decimal expansions. For example, in the case of $\frac{2010}{5010} = \frac{201}{501}$, the decimal approximations rounded to 11 decimal places are 0.40119760479. However, because the decimal continues beyond what is shown, we cannot conclude that the two fractions are equivalent merely by comparing the decimals (their decimal equivalents might differ in later places). For example, $\frac{3333333333}{9999999999}$ and $\frac{3333333333}{10000000000}$ both have the decimal approximation 0.3333333333 when rounded to 10 decimal places, but they are not equivalent fractions.

Participants may be tempted to incorrectly cancel the zeroes in $\frac{2001}{6003}$ and $\frac{204}{408}$. Indeed, one can prove that $\frac{2001}{6003}$ is equal to $\frac{21}{63}$ by showing that both are equivalent to $\frac{1}{3}$ and that $\frac{204}{408}$ is equal to $\frac{24}{48}$ by showing that both are equivalent to $\frac{1}{2}$. However, the "canceling" of middle zeroes in these cases is not connected to division by a factor of a multiple of 10 (the way it is in the case of $\frac{6200}{8700}$). The "canceling" works in these cases because the smaller numerals are the result of dividing the numerator and denominator by the same number, a common factor of both. Therefore, the ratio of the new numerator to the new denominator is the same as in the original. These special cases share a commonality with fractions whose numerator and denominator end in zeroes. When the numerator and denominator of these special case fractions are divided by a common factor, the resulting equivalent fraction loses its zeroes. Yet, because the result looks like simple removal of interior zero digits, these problems are potentially confusing.

In Mr. Punzak's alternate problem of $\frac{101}{201}$, because 101 and 201 have no factors in common, this fraction cannot be simplified at all and therefore "canceling" zeroes is not an option (in fact, one can see that $\frac{11}{21} > \frac{101}{201}$ by looking at their decimal approximations on a calculator). It is important to note, however, that in a similar problem, for example $\frac{102}{202}$, the numerator and denominator do have a common factor, 2, but the interior zeroes cannot be "canceled" because $\frac{102}{202} \neq \frac{12}{22}$, the fractions have different values.

What is the mathematics behind those fractions for which "canceling" does not produce a fraction that is equivalent to the original one?
Here participants may offer that for these fractions, the numerator and denominator cannot be divided by the same number, that no common factors exist between numerator and denominator, or that each fraction is represented by a different decimal number.

What are some rules you came up with about "canceling"? Which of these rules apply to all of the fractions on the homework sheet? How confident are you that these rules will apply to all fractions with zeroes in the numerator and denominator?
These questions about rules go to the heart of the discussion that takes place in the videotaped clip. The ability to come up with a rule or a generalization from specific cases is an important one for students to develop in mathematics classes. Participants should be aware that this is the kind of thinking that is expected of students at this level.

Participants are likely to come up with a variety of rules for "canceling" zeroes. Here are a few examples.

1. If zeroes come at the end of the numerator and denominator, they can be "canceled."
2. If zeroes come in the middle of the numerator and denominator and the numerator and denominator are multiples of each other, the zeroes can be "canceled," as in $\frac{204}{408} = \frac{24}{48}$. (If participants suggest this rule, you might want to point out that it works because the ratio of 204 : 408 is the same as the ratio of 24 : 48.)
3. If the decimal equivalent for a fraction with a zero in the numerator and in the denominator is the same before and after crossing out the zeroes, then the zeroes can be "canceled."
4. If the numerator and denominator both have a common factor of 10, the zeroes from the end of the fraction can be "canceled."

ACTIVITY 2
Video
45 minutes

> ***Materials***
> Videotaped Clip 6
> Handouts 31 & 32

Observing the "Canceling" Zeroes Video

15 minutes

First viewing of "*Canceling*" *Zeroes* Let participants know that they will view twice a 12-minute videotaped episode from a fifth classroom. The teacher, Mr. Punzak, asks his students to explore the question of when it is O.K. to "cancel" zeroes, given a fraction with zeroes in both the numerator and the denominator.

> The intent of this first viewing of the videotape is to give participants practice using their observation skills by focusing on one section of the Observation Guide only. At the same time, when groups share their findings, participants will gain a more complete picture of what is salient about this classroom episode.

Distribute Handout 31, the Overall Observation Guide. Organize the participants into three groups: one for each of the three sections of the Guide: Math Content, Learning & Pedagogy, Intellectual Community.

Suggest to participants that they take notes as they view the clip. Give participants a few minutes to finish their notes after they have finished viewing the clip.

15 minutes

Sharing Observations about "*Canceling*" *Zeroes* Focusing on each section of the Overall Observation Guide in turn, ask participants to share their observations. You might ask:

What stood out for you as you were viewing this video?

As participants make their comments, write them in the appropriate sections on the large guide you prepared ahead of time. Encourage participants to connect their comments with what they observed in the classroom and to stay away from generalizations as much as possible. Ask participants to be respectful of the students and Mr. Punzak in their comments, to cite specific evidence for points they make, and to form conjectures about events that puzzled or intrigued them.

Because time is limited for this first viewing, you may wish to defer further discussion of participants' observations until after the second viewing, when you will delve more deeply into the mathematical ideas that students are sharing and the ways the teacher works with them.

15 minutes

Second viewing of "*Canceling*" *Zeroes* Distribute Handout 32, the Learning & Pedagogy Observation Guide, and tell participants that they will be using it during the second viewing.

Remind participants that the bulleted questions encompass facets of the boldfaced questions and are designed to provide participants with concrete ways to identify what the boldfaced questions address.

Read through and discuss the questions on the Learning & Pedagogy Observation Guide to make sure that participants understand what each of the questions is asking. Because some participants may be uncomfortable with the idea of students being confused in a mathematics class, you might elaborate a bit as you explain the intent of this set of questions. You might point out that often teachers and administrators think of confusion as an indication of deficient thinking in need of being fixed; however, in this course, we view confusion as an indication of fragile thinking at the edge of what a student already understands and is working hard to grasp. You might also talk about confusion as a naturally occurring, albeit transient, state in classrooms in which challenging mathematical ideas are being explored. When teachers work productively with students, confusion can lead to new thinking.

Before you show the video a second time, suggest to participants that they take notes using the questions on the guide. After they have finished viewing the episode, allow participants a few minutes to finish up their note taking. Mention to participants that although they will not be discussing their observations of the class, the attention they pay to students' mathematical ideas and the way the teacher works with them will inform the transcript analysis they will do next.

Synopsis of the Classroom Episode

This videotaped classroom episode was chosen for the way it shows a teacher working with students' mathematical ideas, many partially formed, and using these ideas to drive instruction.

Mr. Punzak rethinks his plan for the day. The episode begins with a student doing a division problem on the board that has a remainder of $\frac{10}{30}$, which she expresses as $\frac{1}{3}$. When Mr. Punzak (the teacher) asks students to explain where the $\frac{1}{3}$ comes from, one student says, "Just take out the zeroes," and another, "You got to 'cancel' the zeroes." Mr. Punzak decides that instead of following the lesson he had originally planned that day, he will explore their understanding of "canceling" zeroes further.

Mr. Punzak poses an alternate problem to challenge their thinking (called a counterexample in the transcript): "Could you 'cancel' the zeroes in $\frac{10}{101}$ and call it $\frac{1}{11}$?" Mr. Punzak poses this problem to push his students' thinking about when it is O.K. to take out or "cancel" zeroes in a fraction. Students comment that the zeroes have to be in the same column (i.e., have the same place value) and the numerator has to go into the denominator to be able to "cancel" zeroes. Mr. Punzak decides to follow the idea that the zeroes have to be in the same column, so he poses a second alternate problem.

Second alternate problem:
"Can you 'cancel' the zeroes in $\frac{101}{201}$?"

Students raise the following points:

> You can't because 11 doesn't go into 21.
>
> The zeroes have to be in the same column, but they also have to be in the far right.

Joe poses his own problem:

> What happens if you have two zeroes next to each other as in $\frac{200}{700}$? Can you "just make an X" and cross all of the zeroes out?

Mr. Punzak asks again whether the zeroes in $\frac{101}{201}$ can be "canceled." Several girls are talking back and forth to each other about their ideas. One girl, Kate, concludes that $\frac{11}{21}$ equals $\frac{101}{201}$. Mr. Punzak goes back to the board to pose the problem again. Another girl, Tira, explains that $\frac{101}{201}$ is equal to $\frac{11}{21}$ because both are just shy of being *half* of either $\frac{101}{202}$ or $\frac{11}{22}$. She asks Mr. Punzak whether he thinks she is right and he tells her he is not going to tell her what he thinks.

Students discuss "canceling" the ones in $\frac{101}{201}$. Lisan concludes that "you can't 'cancel' the ones, but you can 'cancel' the zeroes." Mr. Punzak asks Lisan whether she is willing to accept that by canceling the zeroes, she is saying that $\frac{101}{201}$ is equal to $\frac{11}{21}$. This is followed by more discussion about "canceling" the ones after which the discussion moves back to the zeroes. Tira concludes that it is O.K. to "cancel" the zeroes because the zeroes are placeholders. Mr. Punzak asks the class whether everyone is happy with "canceling" out the middle zeroes. Kate replies that because 11 goes into 21 the same as 101 goes into 201 she thinks it is O.K. to "cancel" the zeroes.

Nora's insight. Nora follows up by saying that to know whether it is O.K. to "cancel" zeroes, they need to divide 101 by 201. Kate does this division and 11 divided by 21 as well and finds that the two answers are different. Mr. Punzak reiterates that the two fractions are different and asks Joe what that tells him. Joe's reply is inaudible. Mr. Punzak says that the two fractions are not the same ($\frac{101}{201}$ does not equal $\frac{11}{21}$), so the zeroes cannot be "canceled."

Mr. Punzak's homework assignment.
Mr. Punzak reminds his students of the prime factoring they did when they found out about "canceling." He tells them that prime factoring is a clue to why the zeroes in $\frac{101}{201}$ cannot be "canceled." Their homework assignment is to use the idea of prime factoring to figure out why the zeroes in $\frac{101}{201}$ cannot be "canceled."

Mathematical Highlights of the Tape

The Use of Alternate Problems
Mr. Punzak structures the class around the posing of a series of related questions that address and challenge the rules that students generate. These problems bear some similarity to the original case but are different in certain critical ways. They force students to reconsider what they think they know. For example, using the case of $\frac{10}{30}$, for which students are comfortable with the idea that "canceling" works, Mr. Punzak poses a new fraction, $\frac{10}{101}$, where the reasoning they used to figure out the previous case is challenged. As the class progresses, Mr. Punzak poses another problem $\frac{101}{201}$, and then a student poses his own, $\frac{200}{700}$. Each time students are forced to reconsider what they believe to be true about the "canceling" zeroes rule.

Determining equivalency
Students are also grappling with how to tell whether the original fraction and the one from which the zeroes were "canceled" are the same. Students use different approaches to determining equivalence. An example of this is the part-whole reasoning Tira uses early on in the episode to explain how she knows that $\frac{10}{30}$ is equal to $\frac{1}{3}$. She says:

> When you divide something into thirds, that you're putting it in three equal parts, and in thirty, ten is one of the three equal parts. Ten plus ten plus ten equals thirty. And one plus one plus one equals three.

This is an indication of at least a partial understanding of fractions as representing portions of a whole and of the relationship between the numerator and the denominator in fractions (i.e., that $\frac{10}{30}$ is $\frac{1}{3}$ of $\frac{30}{30}$, or a whole, in the same way that $\frac{1}{3}$ is $\frac{1}{3}$ of $\frac{3}{3}$, or a whole). It might have been interesting to explore this student's thinking further, both to support her articulation of these emergent ideas and to make them accessible to the rest of the students.

Another student proposes to compare the distance of each fraction from half to determine equivalence. (See the discussion of this in **Synopsis of the Classroom Episode.**) Although this student's thinking was off, this could be a mathematically valid way to think about the problem. In order to use a valid method, the student could measure the "distance from half" differently to consider the value of the fraction and compare that value with half. For example, fractions $\frac{1}{8}$ larger and $\frac{2}{16}$ larger than $\frac{1}{2}$ would be equivalent because $\frac{1}{8}$ and $\frac{2}{16}$ are equivalent. Again, it might have been interesting to pursue this idea further.

Students also propose comparing decimal equivalents of both the original and the "canceled"-zeroes fraction to determine equivalence. They eventually determine that $\frac{101}{201}$ is not equivalent to $\frac{11}{21}$ because the decimal equivalent of $\frac{101}{201}$ is approximately 0.5025 and the decimal equivalent of $\frac{11}{21}$ is approximately 0.5238.
The other students seem to accept this conclusion on faith. It would have been interesting to explore the ways in which they made sense of the process of dividing the numerator by the denominator in order to determine the equivalent decimal number. What are these students' understandings of how fractions represent the operation of division?

In general, this videotaped clip raises a number of questions about the students' understanding of equivalent fractions. How do these students understand fractions as representing parts of wholes? What are their understandings of the relationships between the numerators and the denominators of equivalent fractions? These ideas about fractions—parts of wholes and the relationship between numerators and denominators in equivalent fractions—are some of the ideas behind rules about and procedures for "canceling" zeroes that are important for students of this age to explore.

Throughout this videotaped clip, we see Mr. Punzak being open to all of the students' contributions. He chooses to pursue some of the students' ideas while letting others go by. He refrains from divulging answers or voicing his opinion on whether a certain way of thinking is on the right track. Not until the end of the class when he is assigning homework does he ask students to think about the problem in a particular way.

We are left with a number of questions about students' understanding of the mathematics in this lesson and about the degree to which Mr. Punzak understands how they are thinking mathematically. This leaves room for participants in your group to conjecture about how individual students are thinking about the ideas of equivalent fractions, factors of 10, and prime factoring.

Threaded through the videotaped episode we see many examples of students engaged in mathematical sense making, responding to Mr. Punzak by formulating and defending their positions, posing and testing hypotheses, and generating some of their own problems. We see them listening to one another and responding to one another's ideas.

First Viewing of the Classroom Episode

For the first viewing, participants will use the Overall Observation Guide. One third of the participants will concentrate on the Intellectual Community section, one third on the Math Content section, and one third on the Learning & Pedagogy section.

Sharing Observations

For this part of the activity, you will record participants' observations in the appropriate columns of your prepared chart. It may be interesting for you to notice whether participants' observations, especially those about the intellectual community and math content of the lesson, are more detailed than they were in earlier sessions of the course when these ideas were newer.

Some participants may observe that Mr. Punzak does almost no "direct teaching" during this class and allows his students to leave the class without a firm understanding of the problem they have been grappling with. Lack of closure by the end of a class is a big issue for many administrators who are uncomfortable with the idea of students leaving a class not knowing the correct answer or at least how to obtain it. Administrators who think this way may not appreciate the complexity of the ideas students are grappling with, and they may not be considering that students may need more time than is available in a single class period to work their way through such complex mathematical ideas.

Other participants may comment on the nonjudgmental way in which Mr. Punzak responds to students' ideas and students' willingness to take intellectual risks in this classroom.

Second Viewing of the Classroom Episode

For this viewing, participants use the Learning & Pedagogy Observation Guide. It is important to give participants a few minutes to look over the Guide in some detail and to walk through it with them.

The Questions in the Guide

The questions as a whole reflect the view that students' ideas are in progress. They may have a fragile understanding of a concept at a particular point in time—a partially formed idea, or an idea that seems to be poorly understood—but as they work to develop their thinking, their understanding becomes increasingly more robust. This is in contrast to the view that students either "get" a concept or they do not, and if they do not, their thinking is deficient in some way. The challenge for teachers is to figure out what a student understands and what he or she still needs to work on, and to use these insights to decide what would be the most productive way(s) to help him or her further develop ideas.

The first set of questions—**What kinds of mathematical sense-making are students doing?** and **How does the teacher work with the sense the children are making?**—is meant to draw the observer's attention to the core of a mathematics class: students' ideas about the mathematics and the response of the teacher to these ideas. What the teacher does to work with students' ideas is critical to students' development as mathematical thinkers and is at the heart of this session as well as the next one. Pushing students' thinking by posing alternate problems and probing certain ideas while letting others pass by are some of the ways teachers can work with students' ideas.

The second set of questions—**What mathematical ideas seem to be confusing to students?** and **How does the teacher work with students' confusions?**—highlights the idea that confusion is a naturally occurring, albeit transient, state in classrooms in which challenging mathematical ideas are being explored and that when teachers work productively with students, confusion can lead to new thinking. In other words, confusion is part of the journey toward understanding.

Often teachers and administrators think of confusion as an indication of thinking that is deficient in some way and that needs to be fixed. Participants who are not comfortable with the idea of confusion may not appreciate that students' understandings of a mathematical concept will be fragile while they are at the edges of what they understand well and what they have not yet grasped and that students need time to make sense of the ideas for themselves. Posing alternate problems or helping students see how the idea they are working on is connected to a related one they have worked on before are some of the ways teachers can work productively with students' confusion.

The third set of questions—**How are students developing their mathematical ideas over time?** and **How does the teacher show that he or she understands that students' ideas develop over time?**—addresses the ideas that mathematical thinking develops gradually and that at any one time, students in the class are likely to have a range of ideas about a particular concept. Furthermore, individual students are likely to understand some of the ideas they are exploring more fully than other ideas; at the same time, each student's understanding is in flux as he or she works towards deeper and more complete comprehension. It is often difficult to find evidence of students' ideas developing over time in one class period because the time frame of one class is so short.

The last two questions in the Pedagogy column have no counterparts in the Learning column. The first—**How does the teacher appear to adjust his or her teaching based on the ideas that he or she hears from students?**—underscores the close connection between students' mathematical thinking and the teacher's pedagogical moves.

The final question—**How does the teacher attend to all students?**—addresses the idea that heterogeneity exists in all classrooms and that how the teacher attends to diversity along multiple dimensions such as learning styles, background knowledge, primary language, and individual cognitive strengths and weaknesses must be taken into consideration when observing in classrooms.

Participants will not discuss their observations of the classroom episode after the second viewing of the videotape. However, the observations they make using the questions on the Learning & Pedagogy Observation Guide will inform the work they do next, analyzing students' mathematical thinking and the way(s) the teacher works with their ideas.

ACTIVITY 3
Small Group Analysis
45 minutes

Materials
Chart paper and markers
Handout 29
Reading 8
Reading 3

Making Sense of the Learning and Pedagogy in Mr. Punzak's Classroom

20 minutes

Working with the Transcript Explain to participants that they will be examining some of the pedagogical moves Mr. Punzak made in response to the mathematical sense making of his students in order to understand in a grounded way what is meant by some of the questions and terms on the Learning & Pedagogy Observation Guide. Tell them that they will be working with one section of the transcript as they do this.

Distribute the Guide, and ask participants to form groups of three or four. Distribute Handout 29, Working with the *"Canceling" Zeroes* Transcript, to each group. Tell participants to open to the transcript of *"Canceling" Zeroes* (p. 113 in the book of *Readings*).

Groups will work with one of the two underlined sections of the transcript (lines 55–73 or 129–149). You may want to assign groups to sections to ensure a balanced distribution. The groups are to analyze the section of the transcript, using the bulleted questions on the Learning & Pedagogy Observation Guide. Remind them to focus on the interplay between students' mathematical thinking and the way Mr. Punzak works with their ideas.

Participants are also to flag those places in the transcript where students' thinking reflects partially understood ideas or ideas about which they are confused. Participants will consider these instances of students' thinking during the next activity, when they develop questions about students' mathematical thinking that they would like to probe further with Mr. Punzak.

Mention to participants that in the discussion following this activity, they will be sharing their ideas about the learning and teaching taking place in this class, and will suggest ways to address each of the questions on the guide.

As groups are working, circulate among them to find out what ideas they are grappling with. Some groups may have questions for you as they begin to use the Learning & Pedagogy Guide.

25 minutes

Whole group

Discussing the Learning and Pedagogy in Mr. Punzak's Classroom Have ready a sheet of chart paper with two columns headed *Students* and *Teacher.*

> Thinking about what questions to ask Mr. Punzak will prepare participants for viewing the videotaped conference between Mr. Punzak and a math supervisor.

Ask participants for their observations about the sense making students were doing and the ways Mr. Punzak worked with their ideas. To ensure that you hear from all groups, you might go from group to group and collect an observation from each.

Make sure that participants focus on the interaction between students' thinking and Mr. Punzak's pedagogical moves. If you find that the groups are not, you might ask them to first contribute observations they have about students' sense making. Then, for each observation, ask for an observation about what Mr. Punzak did in response.

Record participants' observations on the appropriate side of the chart (*Students* or *Teacher*). As often as possible, line up an observation about a specific instance of sense making or confusion on the part of a student across from an observation about how Mr. Punzak worked with it.

After you have collected data from each group, direct participants' attention to all of the observations, and ask them to think about what they saw in light of the article they reviewed for homework, *What's All This Talk about "Discourse"?* by Deborah Ball (Reading 3).

You might begin this discussion by asking participants the following:

What points does Ms. Ball make about mathematical discourse and the effect of the classroom environment on the quality of that discourse?

What is the nature of the discourse between teacher and students and among the students in Mr. Punzak's and Ms. Ball's classrooms? What do you notice about the ways that both Mr. Punzak and Ms. Ball work with the mathematical sense students are building?

Tell participants that in the next activity, they will be shifting their attention away from the ways that the teacher works with students' mathematical ideas to instances of students' mathematical thinking that participants identified in the transcript. They will use these instances to develop questions to pose to Mr. Punzak in a collaborative inquiry.

Overview

Participants examine some of the pedagogical moves Mr. Punzak made in response to the mathematical sense making of his students. The primary goal is for participants to understand in a grounded way what is meant by some of the questions and terms on the Learning & Pedagogy Observation Guide. The two sections of the transcript were selected because of the way they correspond to many (but not all) of the questions on the Learning & Pedagogy Guide.

It is likely that groups will have time to get only partway through the questions on the Guide as they examine a section of the transcript. Groups are not expected to consider all of the questions on the Guide; rather, the intent of this activity is for participants to tune in to the kinds of questions they should have in their heads and to think about what they need to attend to in the classroom when addressing these questions. They will have more opportunities to work with the Guide for homework and during the next session as well.

From the discussion of each section of the transcript, you can get an idea of the kinds of connections that participants are making between the questions on the Guide and what happened in the classroom episode. Participants may come up with different ideas than the ones presented.

Also in the discussion of the transcript, we point out places where Mr. Punzak might have probed his students' ideas further. We point out these "missed opportunities" in anticipation of the activity that follows, in which participants identify aspects of students' thinking about which they would like to understand more. Participants will use these as jumping-off points for a collaborative inquiry into students' thinking with the teacher.

It is important for participants to keep in mind that teaching is a very complex activity and that all teachers will miss certain opportunities. The purpose of taking a close look at the ways Mr. Punzak works with his students' ideas is not to focus on what he should have done differently, but to do the analytic work of thinking through what his pedagogical decisions reveal about his understanding of his students' mathematical thinking and what more might have been learned through further probing of their ideas.

Summary of Lines 55–73 of the Transcript:

In this section of the transcript, students grapple with whether zeroes can be "canceled" when they are in the tens column in both the numerator and denominator. The discussion reveals partially formed ideas about the mathematics underlying "canceling" zeroes, which moves Mr. Punzak to pose an alternate problem that he thinks might bring them to a new level of understanding.

What kinds of mathematical sense making are students doing?

♦ **What are the partially formed ideas with which students are working?** Students say that the zeroes have to be in the same column, but they do not seem to have a firm enough grasp of the underlying mathematics to understand that "canceling" zeroes is equivalent to dividing by a power of 10 and that thus it works only with numbers that are multiples of 10. Andy (line 55) has an inkling of this when he says that the zeroes have to be in the same column, "like the ones column," but when Mr. Punzak poses the alternate problem of $\frac{101}{201}$, Andy (line 58) indicates that he thinks that "canceling" zeroes would work in this case also. He quickly changes his mind again (line 60) and says that "canceling" zeroes won't work for $\frac{101}{201}$ because 11 doesn't go into 21. His thinking here is an indication of another partially formed idea about the nature of the relationship between the numerator and denominator in order to be able to "cancel" the zeroes.

♦ What ideas are students particularly engaged in thinking about? The general tone of this class is one of serious engagement with the mathematics. In this section of the transcript, an example of this would be when Joe generates his own problem, $\frac{200}{700}$, in response to Nicky's assertion that the zeroes have to be in the ones column.

♦ What ideas seem to be well understood? Students seem to understand that if the zeroes are at the end of the numerator and denominator (i.e., in the ones column), then the zeroes can be "canceled." However, to take these comments as an indication of understanding would be misleading because the students do not seem to grasp the underlying mathematics.

♦ What ideas seem to be poorly understood? There is evidence that some students believe that as long as zeroes are in the same column, they can be "canceled."

How does Mr. Punzak work with the sense the students are making?

♦ What does he do to build on their ideas? Mr. Punzak decides to pursue Nicky's idea that the zeroes have to be in the ones column and to find out how sturdy students' understanding of the mathematics underlying "canceling" zeroes is by posing the alternate problem of $\frac{101}{201}$. By putting the zeroes into the tens column, he pushes students to consider how well Andy's theory that "the two zeroes have to be in the same column" holds up. Through this problem, Mr. Punzak finds that he has indeed found a soft spot in students' understanding of the mathematics that underlie "canceling" zeroes. Having found this soft spot, Mr. Punzak might have benefited from knowing more about what students think is happening mathematically when "canceling" zeroes.

Next, Mr. Punzak instructs students to work in small groups on the problem of whether the zeroes in $\frac{101}{201}$ can be "canceled."

♦ Does he choose not to pursue certain ideas? Mr. Punzak does not pursue the dialogue between Andy and another student about 11 going into 21 (lines 60–66). Again, it might have been helpful to know more about Andy's thinking regarding the nature of the relationship that must exist between numerator and denominator when "canceling" zeroes.

What mathematical ideas seem to be confusing to students?

As was noted earlier, some administrators may find the idea of students being confused in mathematics class unsettling. They may think that a teacher should intervene to clear up their thinking. The questions on the Guide are meant to send a clear message that confusion is to be expected when students are grappling with new and challenging ideas and that understanding can grow out of confusion that is worked with productively.

Although it would be important for a teacher to be able to diagnose what the root of students' confusion might be, administrators would not be expected to have the mathematics background to be able to do this. In this case, an administrator observing a class like this one may decide to pursue the question of what is confusing to students with Mr. Punzak at a post-observation conference.

♦ What might the root of the confusion be? Mr. Punzak asks Nicky why it is true that the zeroes have to be in the far right column (lines 68–73), but he does not push Nicky to articulate why he believes that to be true. We therefore can only conjecture about what the ideas at the root of the confusion might be. In this case, students seem to have lost the connection between "canceling" and the idea that dividing the numerator and denominator of a fraction by the same number maintains the value of the fraction.

♦ What new thinking does the confusion lead to? Mr. Punzak poses an alternate problem of $\frac{101}{201}$, and this seems to push Nicky to decide that in order to "cancel" zeroes, they need to be in the units column. Nicky's assertion then leads to new thinking for Joe who poses an alternate problem with two zeroes in both the numerator and the denominator: one in the units column and one in the tens column (line 75).

How does Mr. Punzak work productively with students' confusions?

♦ **What does he do to help students think through confusing ideas?** Mr. Punzak's carefully chosen alternate problems help push students' thinking to a new level. He appears to be trying to get at Nicky's idea that zeroes have to be in the units place in order to be "canceled" (line 68) when he says, "I wonder why that's true." More information about *why* Nicky believes this—what his understanding of the mathematics underlying "canceling" zeroes is—would have helped Mr. Punzak make subsequent pedagogical decisions.

How are students developing their mathematical ideas over time?

♦ **What changes in thinking do you see that the children are not directly acknowledging?** There is one instance of a student changing his mind about what he thinks. Between lines 56 and 60, Andy initially thinks that it would be O.K. to "cancel" the middle zeroes in $\frac{101}{201}$ and then seems to change his mind right away, saying that "you cannot 'cancel' the zeroes because 11 doesn't go into 21."

How does Mr. Punzak show that he understands that students' ideas develop over time?

♦ **What does he do to help students make connections to prior mathematical ideas?** This does not come up in this section of the transcript.

♦ **What expectations does he have regarding students' levels of understanding, on the basis of the way the session ends?** For most of the class, Mr. Punzak does not seem to be concerned that students come to a particular understanding during the class. Only at the end of the class, after having given students time to explore the ideas on their own, is it clear that one of Mr. Punzak's goals is for students to use prime factoring when deciding whether "canceling" zeroes works.

How does Mr. Punzak adjust his teaching on the basis the ideas that he hears from students?

The first way Mr. Punzak does this is to change his plans for the class session from going over division problems with the class to exploring their ideas about "canceling" zeroes. The rest of the class is a series of pedagogical moves that Mr. Punzak makes in response to the mathematical sense making of his students.

How does Mr. Punzak attend to all students?

It is difficult to get a sense of this during this class session. As you and the participants watch this videotaped classroom episode, you may notice that many of the students who are active participants in the discussion are Caucasian girls. If this issue comes up in the session, it might be helpful to note that the class session has been edited and captures only 15 minutes of a whole lesson. The evident imbalance in participation of students may be an artifact of the editing choices as much as it might be a reflection of the teacher. Point out to participants that for the purposes of this viewing, you would like them to bracket their concerns about this imbalance. Although it is an important issue in elementary mathematics, it is not the focus of this particular session and there is not enough information available to assess it accurately and fairly.

Summary of Lines 129–149 of the Transcript: In this section of the transcript, during which students are still working on whether the zeroes in $\frac{101}{201}$ can be "canceled," the idea of zeroes as placeholders comes up but is not pursued. Then the idea of dividing the numerator by the denominator to check equivalence comes up, and a student uses a calculator to check whether 11 divided by 21 is the same as 101 divided by 201. The student who does the calculations finds that the answers are different, and Mr. Punzak says that this means that the middle zeroes can not be "canceled." Mr. Punzak follows this up with a homework assignment in which he tells students to use the idea of prime factoring to explain why the middle zeroes cannot be "canceled."

What kinds of mathematical sense making are students doing?

♦ **What are the partially formed ideas that students are working with?** Tira puts forth the idea of zeroes as placeholders. This appears to be a reference to a zero holding the place in the tens column in a number like 304 so that the number is read as "three hundred four" (304) instead of "thirty four" (34). Without knowing more about what she meant, it is difficult to know how she understands the idea of zeroes as placeholders.

♦ **What ideas are students particularly engaged in thinking about?** The general tone of this class is one of serious engagement with the mathematics. An example of this would be Nora's excitement about determining equivalence by dividing numerators by denominators to see whether the resulting decimals are the same.

♦ **What ideas seem to be well understood?** Kate and Nora seem to understand that two fractions are equivalent if, when you divide the numerator by the denominator, the resulting decimals are the same. It is not clear how widely shared that understanding is among the students in the class. It would have been interesting if Mr. Punzak had asked why this is the case.

♦ **What ideas seem to be poorly understood?** There is evidence that Kate believes that as long as zeroes are in the same column, they can be "canceled."

How does Mr. Punzak work with the sense the children are making?

♦ **What does Mr. Punzak do to build on their ideas?** Mr. Punzak decides to pursue Nora's suggestion to divide 101 by 201 and 11 by 21 to see whether the two decimals are the same. After it is determined that the resulting decimals are not the same, Mr. Punzak asks what that means.

♦ **Does he choose not to pursue certain ideas?** Mr. Punzak does not pursue Tira's idea of zeroes as placeholders. It might have been useful for him to find out more about Tira's mathematical thinking by pursuing that idea.

What mathematical ideas seem to be confusing to students?

Again, we see that students seem to have lost the connection between "canceling" and the idea that dividing the numerator and denominator of a fraction by the same number maintains the value of the fraction. One also wonders how firm their understanding is of fractions describing parts of a whole.

♦ **What new thinking does the confusion lead to?** After Kate's tentative articulation of how $\frac{101}{201}$ is the same as $\frac{11}{21}$ because "11 goes into 21 the same as 101 goes into 201" (she gets turned around when she sets up these division problems), Nora has her insight about checking the equivalence of the two fractions, suggesting new thinking on her part.

How does the teacher work productively with students' confusions?

♦ **What does the teacher do to help students think through confusing ideas?** After it has been established that 101 divided by 201 does not equal 11 divided by 21, Mr. Punzak tries to move everyone's thinking along by asking what generalization can be drawn from this information. He does not get a clear response to his question, and it is therefore difficult to ascertain how many students are following the mathematics at this point. Mr. Punzak then introduces the idea that prime factors are relevant to the question of "canceling" zeroes.

How are students developing their mathematical ideas over time?

♦ **In what instances do students refer to prior discussions, explorations, or findings?** It becomes apparent that students have had previous experience with prime factors and that Mr. Punzak is bringing the idea of prime factors back in a new context when he suggests that students think about using the prime factors to understand the mathematics of "canceling" zeroes. Because the tape ends at this point, participants cannot know what the effect of his suggestion was on students' understanding of the mathematics.

How does Mr. Punzak show that he understands that students' ideas develop over time?

♦ **What does he do to help students make connections to prior mathematical ideas?** By reintroducing prime factoring into the discussion, Mr. Punzak is helping students use an idea they encountered earlier. Mr. Punzak is also showing that he understands that students may not fully understand a concept the first time they encounter it.

♦ **What expectations does the teacher have regarding students' levels of understanding, given the way the session ends?** Mr. Punzak gives students the time to explore and do their own thinking during this class. When class ends, he seems comfortable with the idea that class is ending before students have become clear about the mathematics underlying "canceling" zeroes. This indicates that Mr. Punzak acknowledges that students' understandings of the ideas deepen over time and that they need more time to work through the ideas. The hint he provides about prime factoring at the end of the class may indicate that he thinks that students may need another level of intervention to move their thinking along.

Mathematical Ideas to Probe Further

As participants are examining the sections of the transcript, they will be looking for places where student thinking reflects partially understood ideas or ideas about which they are confused. These are examples of student thinking participants might want to flag for further consideration during the next activity, when they develop questions about students' mathematical thinking that they would like to probe further with Mr. Punzak. An example of one such place might be in line 60 when Andy said that the zeroes in $\frac{101}{201}$ cannot be "canceled" because 11 doesn't go into 21. Here Andy reveals a little piece of his thinking. Probing it further would have made it possible to develop a fuller understanding of his ideas, information which a teacher could use to make pedagogical decisions. Another example is in line 130 when Tira refers to zeroes as placeholders. Again, a fuller understanding of her ideas here might be of use to a teacher as he builds on his students' thinking.

Discussion of the Learning and Pedagogy in Mr. Punzak's Classroom

The Ball article *What's All This Talk About "Discourse"?* that participants revisited for homework provides a good framework for thinking about the interplay between Mr. Punzak and his students around mathematical ideas. Ms. Ball points out that NCTM's *Professional Teaching Standards* emphasize the importance of discourse in mathematics classrooms. She encourages the use of this document as a tool to promote constructive discussions about teaching. She includes an extended example of a discussion that took place in her third grade classroom with annotations that include the questions she asked herself that led to the pedagogical decisions she made.

Discussion Questions

♦ **What points does Ms. Ball make about mathematical discourse and the effect of the classroom environment on the quality of that discourse?** This question asks participants to think about the interaction between the mathematical ideas at play in a classroom and the intellectual environment of the class. Ms. Ball describes mathematical discourse as a way for knowledge to be "constructed" and "exchanged" in the classroom. Although discourse is formed by both students and the teacher, Ball insists that teachers play an important role in shaping it. She warns that if the discourse patterns are left unattended, they can revert to long-established norms such as competitiveness among students, the teacher as the source of answers, and an emphasis on one right way of working out a problem.

Ms. Ball asserts that the classroom environment can affect the quality of the discourse that takes place. She lists factors such as how safe students feel, how respectful they are of each other's contributions, to what extent the mathematical thinking is considered important, and how much the teacher values students' contributions.

♦ **What is the nature of the discourse between teacher and students and among students in Mr. Punzak's and Ms. Ball's classrooms? What do you notice about the ways both Mr. Punzak and Ms. Ball work with the mathematical sense students are building?** What stands out in both classrooms is the centrality of students' voices and their serious engagement in mathematical thinking. The patterns of discourse in both settings show many student-to-student dialogues. On the other hand, the teachers' voices in both settings are a serious presence as each teacher helps shape the discourse. In Mr. Punzak's case, he poses alternate problems, structures and paces the work that is done in small- and whole-group formats, repeats students' ideas, and asks others to respond. Ms. Ball's shaping of the discourse takes the form of asking for other students' reactions to an idea a student posed, keeping the discussion on track, and pushing students' thinking by asking why something a student suggested might be a good idea. Students in both settings seem comfortable sharing their ideas; they listen to one another and respond to one another's ideas.

ACTIVITY 4
Whole Group Discussion
45 minutes

> ***Materials***
> Chart paper and markers
> Videotape 7
> Handout 30
> Reading 8

An Image of Collaborative Inquiry

15 minutes

Developing Questions for Collaborative Inquiry
Tell participants that they will use the examples of students' thinking they identified in the previous activity to come up with questions about students' mathematical thinking with which to engage in collaborative inquiry with Mr. Punzak. You might begin by asking participants for instances of partially formed or confusing ideas they identified. Ask them to develop a question about students' thinking that would invite collaborative inquiry with the teacher.

> The intent here is for participants to gain experience forming collaborative inquiry questions to prepare them for their viewing of the videotaped interview. Participants also practice developing collaborative inquiry questions before they do their next homework assignment in which they come up with their own questions to ask during a post-observation conference with a teacher they have just observed.

Keep a list of the questions participants pose on a large sheet of chart paper. After you record each question, ask participants to look it over for tone and substance. Ask them whether the question is about students' mathematical thinking and whether it reflects a collaborative stance that invites someone to enter into a conversation between colleagues. If appropriate, you may want to solicit wording changes to the participants' questions.

Let participants know that next they will see an actual case of collaborative inquiry as they watch a videotaped conference between a teacher and a math specialist.

15 minutes

Viewing the Videotaped Clip Before you show the videotaped clip, provide participants with some background about it. Let them know the following:

- The conference is between Mr. Punzak, from the *"Canceling" Zeroes* episode, and Ms. Davidson, a mathematics specialist.
- The interview was made especially for this course to model what takes place during a conference in which the two parties are investigating together the mathematical thinking that took place during a class.
- The 14-minute interview is from an hour-long conference between Mr. Punzak and Ms. Davidson during which they addressed a wider set of questions than the subset shown on the tape.
- Although collaborative inquiry is used throughout the conference, Ms. Davidson also uses other modes of communication.

Remind participants that they will view the interview only once. Distribute Handout 30, Reflecting on Co-Inquiry, and have participants look it over before they watch the interview. This handout lists the discussion questions participants will address following the viewing of the tape.

What stood out for you as you watched the videotape?

How might a conversation between a teacher and an administrator who is not a math specialist differ from the one that takes place on the videotaped clip?

How might this conference have supported Mr. Punzak's generative learning?

Suggest to participants that as they watch the video, they may want to take notes on the questions Ms. Davidson asks and the ensuing conversation.

15 minutes
Whole group

Discussing the Videotaped Clip Following the viewing of the videotaped interview, have a whole-group discussion of the questions that are listed on the handout.

Overview

As has been discussed at earlier points in this course, collaborative inquiry into students' mathematical thinking is a new idea for many administrators and requires a shift in the way they usually talk to teachers. Instead of thinking that they need to provide teachers with a set of conclusions about their classroom practices, administrators using a collaborative inquiry approach come into a conference ready to have a conversation with the teacher in which both discuss the sense they got of what students were thinking and both decide on aspects of students' thinking that they would like to probe further. When administrators engage in these kinds of conversations with teachers, they emphasize that being curious about and understanding students' mathematical thinking are central. Research shows that being curious about students' mathematical thinking is the single most important factor in developing into a teacher who continues to learn.

Whole Group Development of Questions

Participants will use the examples of students' thinking they identified in their transcript analysis to develop questions that they would be interested in co-inquiring about with Mr. Punzak. It will be important here to encourage participants to look over the questions being posed to make sure that they are about the students' mathematical thinking and that they set an inviting tone that would be characteristic of a conversation between colleagues.

A promising collaborative inquiry question into student thinking grows out of a genuine curiosity that the questioner has about an idea expressed by a student during the class. The question often takes an open form and can be posed by either person. Some examples might include the following:

> I was really interested in and puzzled by what Tira meant when she said that zeroes were placeholders. I wonder what was going on there?

> That discussion about "canceling" the zeroes in $\frac{101}{201}$ really surprised me. I wonder what "canceling" means to those students?

Viewing the Videotaped Conference

This 14-minute videotaped clip shows parts of a conference between Mr. Punzak and Ms. Davidson, a mathematics specialist. It was made especially for this course for the purpose of modeling what a conference in which the two parties are investigating together the mathematical thinking that took place during the class would look like. The entire conference covered a range of topics in addition to students' mathematical thinking, from number theory to the distinction between proof and evidence to the intellectual culture of the class.

Before participants view the videotape, let them know that the intention here is to provide them with an image of what collaborative inquiry into students' mathematical thinking looks like rather than for them to follow in detail the mathematics Mr. Punzak and Ms. Davidson discuss.

Summary of the Conference

The abridged version of the conference begins with Ms. Davidson asking Mr. Punzak what was behind his decision to change his plan for the day and what he was hoping students would learn mathematically from exploring the topic of "canceling" zeroes. Mr. Punzak explains that he shifted his plan because "canceling" zeroes was a relevant topic and students' interest in it was high. He adds that he wanted to see whether students would use prime factoring, a topic that they had worked with before in the context of finding equivalent fractions, as a way of understanding the mathematics underlying "canceling" zeroes. He explains that students always want to know the rules—they are looking for procedures—but he is interested in their understanding the reasons for the rules. He agrees with Ms. Davidson that there was a fair amount of sense making taking place in the class; however, he says that he didn't feel that the students succeeded in making sense of the underlying mathematics.

Ms. Davidson probes Mr. Punzak's thinking further, asking what he wanted students to understand in terms of prime factoring. Mr. Punzak explains that he hoped one of the students would say, in the case of $\frac{101}{201}$, that 101 is prime and 201 is not and that because they have no factors in common they cannot be reduced. Ms. Davidson then suggests that they try to make sense of some of the students' thinking that took place during the class session.

First Example of Students' Thinking. The first example of student thinking Ms. Davidson brings up is Andy's. Andy said that the zeroes from $\frac{101}{201}$ cannot be "canceled" "because 11 does not go into 21." Mr. Punzak initially says that he is not sure what Andy meant and then proposes that "maybe he's saying that 11 and 21 don't stand in the same relation as 101 and 201," and that if 101 had been 102, then "canceling" the zeroes would have been O.K. Ms. Davidson suggests that if 101 had been 102, Andy would likely have "canceled" the middle zeroes, thus proving to himself that it is O.K. to "cancel" zeroes in the tens column. Mr. Punzak agrees that this is what Andy might have done and says that he, Mr. Punzak, would then have talked about the difference between evidence and proof.

Ms. Davidson suggests that they look back together at what other students were saying and try to understand their thinking. Mr. Punzak says that he is not sure how well he will be able to "unriddle" their thinking, but he agrees to try.

Second example of students' thinking. Next, Ms. Davidson brings up Nicky's comment that the zeroes have to be in the same column and that they need to be in the far right one. Using a new example, $\frac{270}{970}$, Ms. Davidson asks Mr. Punzak what he thinks would help Nicky understand more fully the mathematics of zeroes in the unit column. Mr. Punzak conjectures that if he had asked Nicky about "canceling" the sevens in $\frac{27}{37}$, Nicky might have said that it would be O.K. What Mr. Punzak doesn't say, but is implied, is that Nicky would have realized that if you "cancel" out the sevens, leaving $\frac{2}{3}$, it would have been clear that $\frac{27}{37}$ was not equivalent to $\frac{2}{3}$ and that this would have helped him understand what is going on mathematically when zeroes are "canceled" in the units column—that "canceling" is related to reducing by a common factor and not merely crossing off digits.

Third example of students' thinking. The third example Ms. Davidson mentions is Tira's comment about "zeroes as placeholders." Ms. Davidson asks what Tira meant when she said that "if you knock out a zero, since it's just a placeholder, it will still mean the same thing." Ms. Davidson first asks, "What's that about?" and follows this question up with another, "And what do you think she understands about zeroes?" Mr. Punzak replies that she seems to think that zeroes can be wedged into the middle of a number. Ms. Davidson then asks him whether he would want to pursue this idea with Tira. Mr. Punzak says he would like to talk to her about place value.

Fourth example of students' thinking. The last example that Ms. Davidson and Mr. Punzak talk about is what he refers to as "the breakthrough moment," when Nora realized that the way to determine whether $\frac{11}{21}$ and $\frac{101}{201}$ are equal is to divide the numerator by the denominator to get a decimal equivalent for each fraction and see whether they are the same. Ms. Davidson checks with Mr. Punzak to see who he thought understood this. Mr. Punzak replies that the students who understood this would be the ones who remember from previous work with fractions that the line separating numerator from denominator indicates division.

Discussion of the Conference

The questions participants discuss after viewing the videotaped conference are discussed below.

♦ **What stood out for you as you watched the videotape?** Participants may note the open-ended nature of Ms. Davidson's questions and her focusing of Mr. Punzak's attention on student understanding. They may also note that because only a small part of students' mathematical thinking was uncovered during the class, Mr. Punzak and Ms. Davidson could only conjecture about what students might have meant by what they said.

If these points do not come up during the course of this discussion, you may wish to mention them yourself as a way of focusing participants' attention on the nature of collaborative inquiry questions. Remind participants that the goal of such an approach is the illumination of students' mathematical thinking.

♦ **How might a conversation between a teacher and an administrator who is not a math specialist differ from the one that takes place on the videotaped clip?** Some administrators may think that they do not know enough mathematics to ask questions about students' mathematical ideas. It may be helpful to mention that in the context of a conversation in which co-inquiry is taking place, it is legitimate to ask questions to which one does not know the answer. It may also be helpful to point out that Ms. Davidson's initial questions tended to be open and of a general nature, such as "What was he thinking when he said that?" or "What's that about?" Although Ms. Davidson may ultimately be able to go more places with the mathematics than an administrator whose mathematics background is not as extensive, the questions an administrator poses can have the effect of stimulating a teacher to find out more about what his or her students are thinking.

♦ **How might this conference have supported Mr. Punzak's generative learning?** Throughout the conference, Ms. Davidson keeps coming back to questions about students' mathematical thinking that Mr. Punzak does his best to make sense of. What becomes apparent is how difficult it is to understand what students were thinking in this class because only a part of their thinking has been revealed. Ms. Davidson's questions point out the importance of teachers' probing students' thinking in order to understand better what students understand and are confused about. An administrator asking similar questions and discussing with a teacher what a student understood could help jump-start a teacher's curiosity into students' thinking that could provide an engaging context for the teacher's generative learning.

CLOSING
20 minutes

Materials
Handouts 33 & 34

Homework

10 minutes

Assigning Homework Let participants know that the homework assignment for the next session is to conduct another classroom observation of one of the two teachers they have selected. They will use the Learning & Pedagogy Observation Guide for this observation. They will conduct both a pre- and a post-observation conference with the teacher. In the post-observation conference, they will try a collaborative inquiry approach as one of the modes of communication with the teacher.

Bridging to Practice

10 minutes

Reflective Writing Remind participants that Bridging to Practice is designed to allow them to focus on the ideas in the course that seem particularly interesting for them and for their school community. Point out that it provides a framework for planning what they can do in order to move themselves and their schools along with these ideas.

Post the following questions, and invite participants to respond to them:

- *Choose an idea that came up today that you found particularly interesting. What is your current thinking about this idea?*
- *Where is your school now with regard to this idea?*
- *What are one or two things that you, as instructional leader, will pursue to move yourself and/or your school along with this idea?*

Your Own Journal Writing

Within one or two days of teaching this class, after you have had the chance to unwind from the class and perhaps talk with a colleague or friend about how it went, set aside some time to write your own journal entry about the class. We encourage you to write about the ideas held by the participants in your group rather than your own actions and sense of how things worked out. In our experience, writing about what the participants are thinking about will be much more useful to you as you prepare for the next session. Some facilitators use the time when administrators are writing in their own journals to make preliminary notes for this journal writing.

Working with the *"Canceling" Zeroes* Transcript

In this activity, you will analyze a section of the transcript in order to understand in a grounded way what is meant by some of the questions and terms on the Learning & Pedagogy Observation Guide.

In groups, you will work with one of the following sections from the transcript:

lines 55–73

lines 129–149

In your small groups, discuss the bulleted questions on the Learning & Pedagogy Guide, focusing on the interplay between students' mathematical thinking and the way Mr. Punzak works with these ideas.

1. Keep track of key ideas from your discussions to share with the whole group.

2. As you examine the transcript, identify those places where student thinking reflects partially understood ideas or ideas about which they seem to be confused. Flag these places for further consideration during the next activity, when you will be developing questions about students' mathematical thinking that you would like to probe further with Mr. Punzak.

Reflecting on Co-Inquiry

As you watch the videotaped conference between Mr. Punzak and Ms. Davidson, keep in the mind the questions below. You will be discussing them after viewing the conference.

1. What stood out for you as you watched the videotape?

2. How might a conversation between a teacher and an administrator who is not a math specialist differ from the one that takes place on the videotaped clip?

3. How might this conference have supported Mr. Punzak's generative learning?

Observation Guide

Students		
Focus Question	Conjectures	Evidence from Classroom
Math Content		
• What mathematical ideas are embedded in the lesson?		
• What makes this worthwhile mathematics?		
Learning		
• What kinds of mathematics sense-making are students doing?		
• What mathematical ideas seem to be confusing to students?		
• In what ways can you see that the students are developing their mathematical ideas over time?		
Intellectual Community		
• How are students showing respect for one another's ideas?		
• How do students use each other as resources as they make sense of mathematical ideas?		
• What evidence beyond raised hands do you have that students are engaged?		

Observation Guide

Teachers		
Evidence from Classroom	Conjectures	Focus Question
		Knowledge of Content
		• What does the teacher seem to understand about the mathematics?
		• What is the teacher's long-term mathematical agenda?
		• What does the teacher seem to understand about the development of children's ideas in this topic?
		Pedagogy
		• How does the teacher work with the sense the children are making?
		• How does the teacher work productively with students' confusion?
		• How does the teacher attend to all students?
		• How does the teacher adjust her teaching based on the ideas she hears from students?
		Facilitating Intellectual Community
		• How does the teacher support students in showing respect for one another's ideas?
		• How does the teacher set the tone so students see each other as resources for mathematical thinking?
		• What interventions does the teacher make to ensure that students' engagement has a focus on mathematical ideas?

Learning & Pedagogy Observation Guide

Students		
Focus Question	**Conjectures**	**Evidence from Classroom**
What kinds of mathematical sense making are students doing?		
• What are the partially formed ideas that students are working with?		
• What ideas are the students particularly engaged in thinking about?		
• What ideas seem to be well understood?		
• What ideas seem to be poorly understood?		
What mathematical ideas seem to be confusing to students?		
• What might the root of the confusion be?		
• What new thinking does the confusion lead to?		
How are students developing their mathematical ideas over time?		
• In what instances do students refer to prior discussion, explorations, or findings?		
• Do students say anything about what they now understand or what they still need to learn?		
• What changes in thinking do you see that the children are not directly acknowledging?		

Learning & Pedagogy Observation Guide

Teachers		
Evidence from Classroom	**Conjectures**	**Focus Question**
		How does the teacher work with the sense the children are making?
		• What opportunities does the teacher give students to share their thinking?
		• What does the teacher do to build on their ideas?
		• Does the teacher choose not to pursue certain ideas?
		How does the teacher work productively with students' confusions?
		• What does the teacher do to help students think through confusing ideas?
		How does the teacher show that he or she understands that students' ideas develop over time?
		• What does the teacher do to help students make connections to prior mathematical ideas?
		• How does the teacher work with students who are in different places with respect to any particular idea?
		• What expectations does the teacher have regarding students' levels of understanding based on the way the session ends?
		How does the teacher adjust her teaching based on the ideas that she hears from students?
		How does the teacher attend to all students?

Homework for Session 7

As administrators, you need to sift through a range of considerations—from factors that relate to the specific teacher's classroom practice to those that concern school and district priorities—as you make decisions about what to discuss with a teacher in a post-observation conference. For this homework assignment, you will focus on the actual mathematical understandings articulated by students: an important focus given that the development of students' mathematical thinking is the goal of good classroom practice. When administrators inquire with teachers about students' mathematical thinking, they are drawing attention to its importance in a classroom and highlighting how critical a detailed understanding of students' mathematical thinking is to the pedagogical decisions teachers need to make. Such a focus on students' ideas lays the groundwork for teachers' growing capacity to engage in generative learning.

Your homework for the next session is to use the Learning & Pedagogy Observation Guide to observe one of the two teachers you have been following for this course. After your observation, conduct a post-observation conference in which you try out a collaborative inquiry approach among the ways you communicate with the teacher, with the idea of supporting the teacher's generative learning in mind.

1. To prepare for this observation, have a pre-observation conference with the teacher, using the Pre-observation Conference Questions.
2. As you conduct your observation and take notes with reference to the Learning & Pedagogy Observation Guide, which you will find in the book of *Readings*, keep track of questions about students' mathematical thinking that occur to you that you might like to ask the teacher during your post-observation conference.
3. Decide what questions and ideas related to student thinking might be fruitful for you and the teacher to investigate together.
4. Conduct a post-observation conference with the teacher, and try out a collaborative inquiry approach where appropriate. Then consider the following:
   a. What ideas about the thinking of students in the class did you and the teacher find especially interesting to explore together? What made them interesting?
   b. What perspectives and additional "wonderings" about student thinking did you come away thinking about? Which do you think the teacher might continue thinking about on his or her own?

Pre-observation Conference Questions

In order to help you make sense of what you will be seeing when you do your classroom observations, plan to meet with the teacher prior to the observation and ask the following questions:

1. What topic will you and your students be working on in this lesson?

2. What do you plan to do in this lesson? (i.e., the origin and structure of the lesson, and so on)

3. What do you hope to accomplish in this lesson?

4. What mathematical ideas are embedded in this lesson?

5. What have you and your students been working on prior to this lesson?

6. How does this lesson fit into your overall goals for the year?

7. Are there students who have special issues in the class?

SESSION 7

Building Mathematics Understanding: More Than the Sharing of Ideas

Given the importance of helping students develop their mathematical thinking, it is critical that teachers structure their lessons in ways that allow them to explore and probe their students' mathematical ideas. However, it is not enough for teachers to simply have students share their initial thoughts; it is also important for them to work with students' ideas in some depth in order to understand the validity of the students' mathematical reasoning. It is when teachers are equipped with such understandings that they can most productively think about next steps to take to pursue and build on students' ideas. This capacity of teachers to continually probe and make sense of their students' thinking and to use what they learn to plan their next pedagogical actions is at the core of generative learning for teachers.

This session provides administrators with their final opportunity of the course to work with videotaped clips on the process of developing their eye for standards-based mathematics classrooms and to further develop their skills in using collaborative inquiry with teachers. These two strands weave their way through the session as participants reflect on the classroom observations and post-observation conferences they did for homework and on the mathematical discourse that takes place in an investigation in a seventh grade classroom. As in the last session, they use the Learning & Pedagogy Observation Guide to focus their viewing.

Overview for Session 7

OPENING page 244 5 minutes	**INTRODUCTION** This is an opportunity to let participants know what they can expect from this session and how it connects to both previous and upcoming sessions.
ACTIVITY 1 Homework Discussion page 246 20 minutes	**EXPLORING GENERATIVE LEARNING** Participants discuss their experiences using the Learning & Pedagogy Observation Guide in their own observations and trying out a collaborative inquiry approach during a post-observation conference with the teacher they observed.
ACTIVITY 2 Mathematics Exploration page 249 30 minutes	**EXPLORING SURFACE AREA AND VOLUME** Participants explore the surface area and volume of rectangular prisms as preparation for viewing the videotaped clip that will follow.
ACTIVITY 3 Video page 252 65 minutes	**DESIGNING PACKAGES VIDEO** Participants view a videotaped clip, *Designing Packages,* in which a seventh grade class investigates the surface area of a number of boxes that have the same volume—24 cubic inches. Participants attend to the students' mathematical ideas, the teacher's approach to working with them, and the questions they would ask the teacher in a post-observation conference.
ACTIVITY 4 Whole group Discussion page 257 50 minutes	**COLLABORATIVE INQUIRY INTO STUDENTS' MATHEMATICAL THINKING** Participants focus on the interplay between students' mathematical ideas and the teacher's pedagogical moves in the *Designing Packages* video. Participants keep the idea of generative learning in mind as they consider what questions about students' mathematical thinking they would be interested in thinking about in a collaborative inquiry with the teacher.
CLOSING page 263 10 minutes	**HOMEWORK** For homework, participants will read Sergiovanni and Starratt's *Sources of Authority for Supervision* (Reading 10) and will write about the homework questions. **BRIDGING TO PRACTICE** Participants finish the session with journal writing that allows them to develop further some of their thinking from the session.

BIG IDEAS

Participants explore:

- the centrality of students' mathematical ideas in a mathematics class
- the role teachers have in shaping the discourse in a mathematics class
- the way administrators' collaborative inquiry with teachers into student thinking supports teachers' generative learning

MATERIALS

- ☐ Handout 33, Homework for Session 7
- ☐ Handout 35, The *Designing Packages* Problem
- ☐ Unifix cubes
- ☐ Graph paper
- ☐ Videotaped clip 8, *Designing Packages* (DVD #3)
- ☐ Handout 36, Working with the *Designing Packages* Transcripts
- ☐ Reading 9, *Designing Packages* Transcripts
- ☐ Handout 37, The Learning & Pedagogy Observation Guide
- ☐ Handout 38, Homework for Session 8
- ☐ Reading 10, *Sources of Authority for Supervision* by Sergiovanni and Starratt

PREPARATION

- Prepare an agenda for the session.
- Write the prompts for Bridging to Practice.
- Think about discussion questions that participants would most benefit from considering, and use these to guide your discussions in all of the activities.
- Do the mathematics assignment yourself for Activity 2. Think about what aspects of this activity participants are likely to find interesting or challenging. This activity works well when participants are in groups of four, with participants with strong math grouped with people in the middle and people with weak math grouped with other people in the middle. This provides a stimulating context for everyone. You might want to think about grouping your participants in this manner for this activity (without telling them the grouping strategy, of course).
- Preview the videotaped clip for Activity 3. Review the guidelines for setting a climate for viewing and discussing the videotaped clip that you will find in the **Introduction.** Prepare a large version of the Overall Observation Guide on a sheet of chart paper to collect participants' observations following the first viewing of the video.
- Work through the excerpts from the video and the transcripts for Activity 4. Mark instances of student thinking to probe further, and then develop some of your own questions about students' mathematical thinking to collaboratively inquire about with the teacher, Mr. Mamer.

OPENING
5 minutes

> ***Materials***
> Name tags
> Agendas

Introduction

Before starting the session, post the agenda in a place where everyone can see it.

5 minutes

Introducing the Session Briefly review the agenda with participants. Let them know that in today's session they will continue to focus on the mathematical sense students are making of the content of a lesson and the ways the teacher works with students' ideas. Participants will do this in the context of discussing both the observations and post-observation conferences they did with the teachers in their schools or districts as well as the classroom episode from Session 6. Participants will once again use the Learning & Pedagogy Observation Guide to focus their viewing. Mention that a parallel focus of the session will be thinking further about engaging in collaborative inquiry with teachers about students' mathematical thinking as a way of supporting teachers' generative learning.

Overview

This is the next-to-last session of the course and a good time to start bringing ideas together. As facilitator, your task is to think about which ideas seem to be most salient to work on with this group and which are most important for the participants to continue working on after the course has ended. In this session, participants continue thinking about learning and pedagogy and their connections to content. The **Facilitator Notes** continue to explicate these connections in addition to exploring the mathematics behind the activity in the videotaped clip. Again, it is important for you as facilitator to use these notes in order to be ready for the range of ideas that participants might raise about the mathematics in the video. The discussion questions are intended to stimulate conversation and to get participants thinking about a set of ideas. Your job is to shape the direction of the conversation and to keep it focused on the central ideas of the session.

Introducing the Session

Participants further develop their eye for standards-based mathematics classrooms in this session as they use the Learning & Pedagogy Observation Guide to observe the mathematical thinking and pedagogical approaches used by the teacher in this session's videotaped clip. They also continue to think about engaging in collaborative inquiry with teachers about students' mathematical thinking as a way to support teachers' generative learning.

ACTIVITY 1
Homework Discussion
20 minutes

Materials
Chart paper
Markers
Handout 33
Handout 37

Exploring Generative Learning

20 minutes
Whole group

Discussing Classsroom Observations Begin the discussion by asking participants to respond to the following questions about the observation they did.

> *What stood out for you as you did the observations using the Learning & Pedagogy Observation Guide?*

> *What was challenging about observing with a focus on students' mathematical ideas?*

The purpose of making sense of students' thinking about the mathematics is to gain insight into both the mathematics *and* students' thinking about the mathematics.

Then turn your attention to the post-observation conference, and ask:

> *What was it like to explore students' thinking with the teacher?*

Probe for some detail about the nature of the exchanges between administrators and teachers so that participants can help one another think through what aspects of this approach seemed to work and what needs further thinking.

Ask for participants' thoughts about keeping the teacher's generative learning in mind.

> *What was the effect of keeping the idea of the teacher's generative learning in mind as you conducted your observation and post-observation conference?*

Overview

For homework, participants were to use a collaborative inquiry approach as one of the modes of communication when talking to the teacher during the post-observation conference. They used this approach to consciously support the generative learning of the teacher they observed as well as their own learning.

As was noted in Handout 32, Homework for Session 7, administrators need to sift through a range of considerations as they make decisions about what issues would be most fruitful to discuss with a teacher, from factors that relate to the specific classroom practices to those that concern school and district priorities. For the purpose of this homework assignment, participants were asked to focus on the mathematical understandings articulated by students. We chose this single-minded focus because it will be a new one for many and because it is such an important idea, given that the development of children's mathematical thinking is the goal of good instructional practice. When administrators inquire with teachers about students' mathematical thinking, they are drawing attention to the importance of mathematical thinking in a classroom, especially in relationship to the pedagogical decisions teachers make. Such a focus on students' ideas lays the groundwork for teachers' growing capacity to engage in generative learning.

For homework, participants were to keep track of questions about students' mathematical thinking that they would be interested in talking to the teacher about and to decide what questions and ideas related to student thinking might be fruitful for them to investigate with the teacher. In addition, they were to conduct a post-observation conference with the teacher in which they try out a collaborative inquiry approach when appropriate. Finally, they reflected on the following two questions:

- **What ideas about the thinking of students in the class did you and the teacher find especially interesting to explore together? What made them interesting?**
- **What perspectives and additional "wonderings" about student thinking did you come away thinking about? Which do you think the teacher might continue thinking about on his or her own?**

Whole-Group Discussion

In a whole-group discussion, participants consider the following questions about the observations they did:

- **What stood out for you as you did the observations and used the Learning & Pedagogy Observation Guide?**
- **What was challenging about observing with a focus on students' mathematical ideas?**

Participants will have a range of reactions to using the Learning & Pedagogy Observation Guide in their schools. For some, the guide may have helped them see things they might have missed previously, such as the following:

- That students seemed to be engaged in the activities, but the activities did not entail much mathematical thinking. Without the Observation Guide, participants may have thought that good mathematics was happening in this classroom because students were engrossed in what they were doing. They may have missed that students were mostly following procedures that did not require much mathematical thinking.
- That a teacher confronts numerous challenges in a classroom when students are in different places in terms of their understanding of the mathematics. Without the Guide, participants may simply have noted that some students seemed confused, and some clearly understood the mathematics. By tuning in to what was clear and what was confusing to students and

how the teacher worked with their mathematical ideas, participants can more easily appreciate what teachers need to be able to do and the mathematics as well as the ideas about students' mathematical thinking they need to understand. Knowing these things can lead to productive thinking on the part of administrators about the kinds of professional development opportunities teachers need to have.

For some participants, the experience of observing with a focus on students' mathematical ideas may have been frustrating because students were not often being asked to articulate their mathematical thinking, which in turn meant that it was possible to observe how the teacher worked with students' mathematical ideas to a limited extent only, if at all. An administrator in this position would not be able to address the questions on the Learning & Pedagogy Observation Guide in a substantive way. In this case, it may be helpful to suggest to participants that a good starting place in doing post-observation conferences with teachers who are not accustomed to asking students to share their thinking might be for the administrator to choose one or two mathematical ideas from the class and share with the teacher his or her own curiosity about what students may have been thinking about these ideas. By sharing their interest in students' thinking with teachers, administrators can draw attention to the important role students' thinking plays in the classroom.

Participants who are not accustomed to focusing on students' mathematical ideas may have found it challenging to attend to students' thinking rather than to the presence or absence of certain forms and structures associated with standards-based mathematics classes such as group work or hands-on activities. Another challenge administrators may identify is difficulty in following students' thinking.

After the discussion about their classroom observations, participants turn their attention to the post-observation conference.

♦ **What was it like to explore students' thinking with the teacher?** Participants' experiences talking with teachers about their students' mathematical thinking are likely to vary greatly. It is important to keep in mind that because this is such a new idea for most administrators *and* teachers, productive conversations are not likely to happen right away. You might spend some time on this question getting into the details of the exchange between participants and teachers so that the group can help one another think through what aspects of the approach they used seemed to work and what needs further thinking.

♦ **What was the effect of keeping the idea of the teacher's generative learning in mind as you conducted your observation and post-observation conference?** Some participants may have noticed a difference in how they experienced the post-observation conference when they kept in mind the idea of the teacher's ongoing learning. Their attention may have shifted from evaluating the teacher to thinking about what they, as classroom observers, can contribute to deepening the teacher's understanding of the students' thinking. Inasmuch as keeping teachers' generative learning in mind calls for putting aside their evaluative hats, some participants may have found this approach difficult. It may take these administrators more time to appreciate how such a focus can provide both teachers and administrators with a context for their own ongoing learning, which will in turn be beneficial to the students with whom they work.

ACTIVITY 2
Mathematics Exploration
30 minutes

> ***Materials***
> Graph paper
> Unifix cubes (enough for each group to have about 50)
> Handout 35

Exploring Surface Area and Volume

30 minutes
Small group

Exploring the Mathematics Let participants know that they will explore the mathematical ideas that a class of seventh graders will investigate on the videotape that they will watch next in this session.

> *You work for the ABC Toy Company, and your job is to find all of the different-shaped packages that could be used to ship 24 cubes (packages should fit the set of cubes exactly—no empty space).*
>
> *Make a chart that shows the length, width, and height for each box with a volume of 24. Make a sketch of each of these boxes, and determine its surface area.*
>
> *How many different boxes are there? How do you know that you have identified them all?*
>
> *Which shaped box uses the least amount of packaging material? Why?*
>
> *Write a formula for calculating the surface area of a box.*

Organize participants into groups of four. Make sure that each group consists either of participants who are strong in math grouped with people in the middle or participants who are weak in math grouped with different people in the middle. Distribute the following to each group: Handout 35, The Designing Packages Problem; graph paper; and unifix cubes.

Let participants know that they have a full half hour to explore this activity. Encourage them to explore it in a way that interests them. There is no need to answer all of the questions.

As participants work in their groups, circulate among them to see how they are approaching the problem and what questions they are raising about the mathematics. This will give you a sense of how far your participants have come mathematically in this course.

Small-Group Work on the Mathematics

The mathematics in this session comes from a *Connected Mathematics* seventh grade unit entitled *Filling and Wrapping* (see Resource list for more information about *Connected Mathematics*). Participants focus on ways to determine the surface area of several rectangular prisms with a fixed volume of 24 blocks and work out which prism shape will have the smallest surface area.

The questions on Handout 34 are not exactly the questions that Mr. Mamer, the teacher, asks his students, but they are designed to give participants the opportunity to explore the main ideas that the students in the videotaped clip explore.

Because this is the last opportunity that participants will have in this course to do a mathematics exploration, it is important that the atmosphere be relaxed and that participants explore the mathematics in ways that interest them so that they leave the course feeling that it is interesting, perhaps even fun, to explore mathematical ideas. For this reason we allow a full half hour for this mathematics with no follow-up discussion. The goal of this activity is for participants to explore enough of the mathematics to understand what they will see on the videotaped clip and to explore mathematics in a collaborative environment. They should not be expected to complete the assignment.

Mathematics Background for This Activity

Rectangular prisms are three-dimensional shapes with six rectangular faces. (They look like boxes.) The *surface area* of the prism is made up of the sum of the areas of all of the faces, calculated in square units. For example, in the rectangular prism in Figure 1, the surface area of the prism is 52 square units. The area of one face is $3 \times 4 = 12$ square units; the surface area of another face is $3 \times 2 = 6$ square units, and the area of a third face is $4 \times 2 = 8$ square units. Because there are six faces, or sides, to the box—two of each shape—the total surface area is $2(12 + 6 + 8) = 52$ square units.

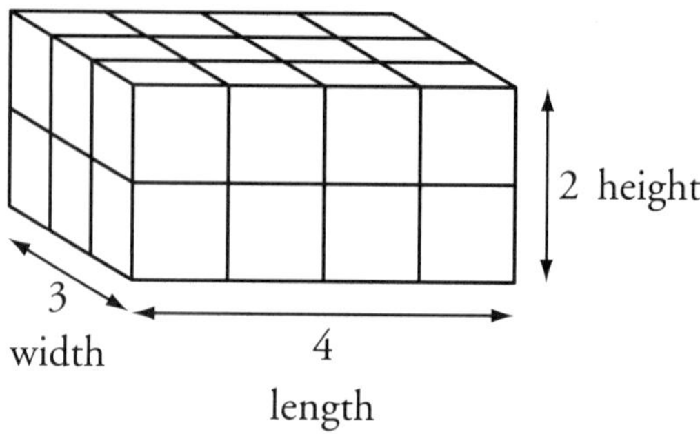

Figure 1

Ways to calculate surface area of the prism include the following:

- Add together the areas of all six faces:
  $12 + 12 + 8 + 8 + 6 + 6 = 52$ square units
- Multiply the areas of like faces by two, and then add:
  $2(12) + 2(8) + 2(6) = 52$ square units
- Add together the area of one of each face, and then multiply by 2:
  $2(12 + 8 + 6) = 52$ square units

Note that these methods demonstrate different ways of representing how the surface area of the prism is composed. Although they look different and represent different ways to think about the surface area, they all refer to the same thing. This equivalence can be demonstrated algebraically as follows:

x = the 12×2 face
y = the 8×2 face
z = the 6×2 face

$$x + x + y + y + z + z = 2x + 2y + 2z$$

$$2x + 2y + 2z = 2(x + y + z)$$

Another way to conceptualize the surface area of a rectangular prism is to look at its *net,* or the surface needed to *wrap* the container. Figure 2 shows the net for the rectangular prism in Figure 1. This net can be folded to become the surface of the rectangular prism, and the area of the flat pattern becomes the surface area of the prism. Looking at the net of the prism provides a way to see surface area as a two-dimensional attribute, even though it is an attribute of a three-dimensional object.

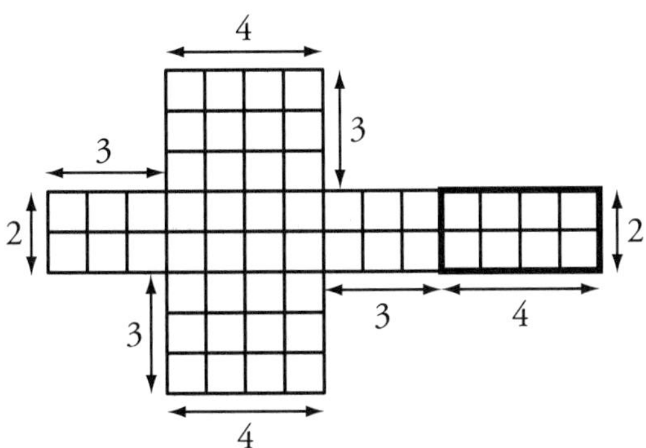

Figure 2

The *volume* of this rectangular prism is 24 cubic units. That is, it would take 24 one-unit cubes to fill this box. This can be calculated by multiplying the length (4) times the width (3) times the height (2).

One of the issues that the students in the video address is the number of ways there are to arrange the cubes in a rectangular prism of a given volume. Of these arrangements, students look for which has the smallest surface area. For the rectangular prism in Figure 1, there are six ways to arrange the cubes that produce different surface areas (see Figure 3). Notice that the length, width, and height of the prism are all factors of its volume (24).

The $4 \times 3 \times 2$ prism has the smallest surface area (52 square units). The closer the prism is to a cube, the smaller the surface area. One way to explain this is that in the cube shape, some unit cubes are fully inside the larger cube, so they have no faces on the exterior face of the larger cube. Other cubes have some interior and some exterior faces. In the arrangements in which one of the dimensions is much larger that the other two (e.g., $12 \times 2 \times 1$), more of the faces of the unit cubes are on one of the surface faces of the larger cube.

Length (in.)	Width (in.)	Height (in.)	Volume (in.³)	Surface area (in.²)	Sketch
24	1	1	24	98	
12	2	1	24	76	
8	3	1	24	70	
6	4	1	24	68	
6	2	2	24	56	
4	3	2	24	52	

Figure 3

You can tell whether you have found all of the possible rectangles with a volume of 24 if you have used all of the factors of 24 in nonrepeating combinations.

It is important to note that some rectangular prisms have the same surface area but the dimensions labeled length, width, or height have been switched (see Figure 4).

Length	**Width**	**Height**	**Volume**	**Surface Area**
4	2	3	24	52
2	4	3	24	52
3	2	4	24	52

Figure 4

These are equivalent rectangular prisms. They are basically the same box at different orientations.

ACTIVITY 3
Video
65 minutes

> ***Materials***
> Handout 37
> Videotaped Clip 8

Designing Packages Video

5 minutes

Introducing the Videotaped Clip Ask participants to have the Learning & Pedagogy Observation Guide in front of them. Suggest that they take notes right on the form as they view the tape.

Remind participants that in an edited segment of a classroom episode such as this one, they may not see evidence of everything included on the Guide. Remind them that they cannot conclude that these elements are missing from the classroom because their data are limited.

> The intent of this first viewing of the video is to let participants react to the classroom episode as a whole before focusing on the details of the learning and teaching that are taking place.

Give participants the following information about the classroom episode:

- The videotape is approximately 30 minutes long.
- It is of a seventh grade classroom doing an activity from a *Connected Mathematics* seventh grade unit entitled *Filling and Wrapping.*

Participants will view the tape in segments, making notes on each segment and discussing the segments immediately after viewing.

Before showing the videotape, remind participants to resist the temptation to make broad judgments about the teacher and the classroom on the basis of the limited view they will get. You might say:

The central purpose for viewing and discussing videos in this course is to stimulate thoughtful discussion among us about the ideas about learning, teaching, and mathematics that are sparked by watching the videotape; it is not to critique the teaching we see. Teaching is complex, and from the perspective of an outside observer there are always other moves that might have been made. Yet an outside observer like any of us, particularly one who sees only a short excerpt of a lesson such as what we will see in the videotaped clip, will not know enough about the background of the class, the context of the lesson, or the teacher's thinking to make valid judgments about whether the teacher's decisions were good ones.

25 minutes

Viewing and Discussing Segment 1 Show Part One Segment One (approximately 15 minutes). After viewing the segment, give participants a few minutes to record their notes, and then begin the discussion. You might ask:

What stood out for you as you were observing this classroom?

Encourage participants to connect the comments they make with what happened in the classroom, staying away from generalizations as much as possible.

20 minutes | **Viewing and Discussing Segment 2** Show Part One Segment Two (approximately 10 minutes). Once again, after viewing the segment, give participants a few minutes to record their notes. Ask again what stood out for them in this segment.

15 minutes | **Viewing and Discussing Segment 3** Show Part One Segment Three (approximately 6 minutes). Then discuss as before.

Overview of the Classroom Episode

This 30-minute videotaped classroom episode was chosen for the images it portrays of a teacher working with students' mathematical ideas. This videotaped clip is part of a series that shows classroom episodes from several NSF-funded middle school mathematics curricula. It has been edited to focus on the teacher's work with his students rather than the curriculum from which it was taken.

Because this tape is long, we have broken it into three segments. Participants will view each segment using the Learning & Pedagogy Observation Guide and will discuss what they see in that segment before moving on to the next. In the next activity, participants go on to view Part 2 of the videotaped clip, which consists of three short excerpts (about 2 minutes each) of the footage they have already seen. These short excerpts, which are accompanied by transcripts, will give participants the opportunity to focus in more depth on how Mr. Mamer, the teacher of this class, works with students' mathematical ideas.

Segment 1 (approximately 15 minutes)

Mr. Mamer reviews previous lessons in which students have discussed the dimensions of a box (or rectangular prism), the number of faces it has, the area of the faces, the surface area of the prism, its net (the amount of material in square units that it would take to wrap the prism), and its volume. Mr. Mamer then presents the task: Students are to design packages that can hold 24 blocks. They will be looking for the most efficient way to find the amount of material it takes to surround the shape.

The students work on this problem in small groups and develop charts on poster paper that describe the surface area and volume of every box they make. Mr. Mamer moves from group to group, discussing the mathematics with the students. The mathematical ideas we see the students exploring include the following: (1) area can be found by multiplying dimensions of a face rather than counting all of the square units; (2) there are different ways to calculate surface area; (3) rectangular prisms with the same dimensions have the same surface area; (4) the volume is constant for all of these prisms; (5) factoring can be one way to find all of the possible rectangular prisms. In addition to exploring these mathematical ideas, students are also working on the skills of organizing and making sense of the data they collect.

One of the shapes that a group constructs contains 24 cubes, but instead of having six flat faces, it has a step like construction along one of its sides. This group discusses whether this particular shape fits the criteria of being a rectangular prism, with a volume of 24 and stackable for transport in a truck. One of the members of the group points out that although it meets the 24-cube and the stackable criteria, it does not have six faces. Students in this group also observe that this shape has the smallest surface area of all the shapes they have made so far.

After this period of small-group work, students post their charts on the wall, and Mr. Mamer leads a whole-class discussion of the patterns the students see in the charts they have posted on the wall. The students' observations include the following: (1) rectangular prisms with the same dimensions have the same surface area, although their orientation may be different; (2) the step-like shape is not a rectangular prism; if it were made into one it would have a volume of 36, which violates one of the criteria; (3) all of the dimensions of all of the rectangular prisms are factors of 24.

Segment 2 (approximately 10 minutes)

The mathematical ideas in this segment are about developing ways of describing how to find the surface area of a prism and evaluating which of these is the most efficient way. Some of the students express their strategies in general terms and use symbolic language, suggesting algebraic reasoning in their thinking.

Mr. Mamer asks students from different groups to share how they wrote the formula for calculating the surface area of the prisms.

The first method:

Area of top $\times 2 = x$
Area of right $\times 2 = y$
Area of front $\times 2 = z$
$x + y + z =$ surface area of prism

The second method:

$6 + 12 + 8 = 26$	(the surface area of three different faces, or half of the surface area of the prism)
$+ 26$	(the surface area of the other half of the prism)
$= 52$	(total surface area)

The third is:

$12 + 6 + 8 + 8 + 6 + 12 = 52$
(the area of each of the six faces)

Mr. Mamer writes these three methods in generalized form on the board.

1. (area of top)2 + (area of right)2 + (area of front)2
2. 2(top + left + front)
3. right + top + front + back + left + bottom

He then asks the students to write these in their notebooks and discuss in their small groups what is similar about those three expressions and what is different about them. The students also talk about which method is easiest and which is quickest. In the whole-group discussion that follows, the students discuss the fact that, although all three methods work, some are easier and faster than others.

Segment 3 (approximately 6 minutes)

When the students made their charts in the first part of the lesson, they did not necessarily order the rectangles in a way that made it easy to see how the surface area of the rectangular prism changed systematically as its shape changed. In the third segment, Mr. Mamer arranges a trip to the library, where the rectangular prisms have been laid out to show what they look like when one dimension has been systematically altered from largest to smallest (see Figure 3 above) so that the students can easily see the relationship between shape and surface area. They return to the class and discuss which rectangular prism has the least surface area and why this is the case. They also discuss the idea that as the shape gets more cubelike, more of the unit cubes have only interior faces, so the surface area of the larger cube is smaller.

Viewing of the Videotape Segment by Segment

Note that, although we usually recommend viewing videotaped clips twice, you will be showing this tape, in toto, only once because of its length. However, in order to make it possible for the participants in your class to think through all of the issues in this long and complex lesson, you will discuss each segment immediately after it is viewed.

For this viewing of the video, participants will use the Learning & Pedagogy Observation Guide. Before showing it, you may wish to refer to the section entitled *Orientation toward Teachers When Observing* in the **Introduction** (p. 10), which is intended to help you think through the purposes of showing and discussing videotaped clips in this course, and consider how you might help participants develop a constructive orientation toward what they see.

It is important to make sure that participants are aware that in an edited classroom episode such as this one, they may not see everything that the Observation Guide asks them to attend to. They would need more information to conclude that such evidence is lacking in a more general way from this teacher's classes.

Sharing Observations about Designing Packages

When you record participants' observations, you may wish to do so on a prepared chart in the appropriate column, as you did in previous sessions. Because your class will be viewing this videotape in segments, their observations will be cumulative. You will be adding comments about later segments to ideas you have already posted on the chart for earlier segments. Participants also may see things in the second or third segments that lead them to change their views of what they saw in prior segments, before the full classroom story was played out. This simulates actual classroom observations, when initial impressions need to be viewed as conjectures, to be confirmed or discredited as the lesson progresses.

ACTIVITY 4
Discussion
50 minutes

Materials
Handouts 36 & 37
Reading 9

Collaborative Inquiry into Students' Mathematical Thinking

30 minutes

Viewing Video Excerpts and Working the Transcripts Ask participants to form groups of three or four. Explain that each group will analyze students' mathematical thinking in one of the three excerpts of the *Designing Packages* video (Reading 9). Assign each group to work on one of the excerpts so that when they are viewing the videotaped clip, they will know which one they will be analyzing more deeply.

Distribute Handouts 36 and 37.

Go over the questions on Handout 36. Let participants know that the first two questions are based on questions from the Learning & Pedagogy Observation Guide that focus on students' partially formed ideas and confusions and how the teacher works with them.

The third and fourth questions are designed to focus participants' attention on students' mathematical thinking, rather than on ways the teacher works with these ideas. Mention this shift in focus and explain the reasoning behind it.

- The idea of engaging in collaborative inquiry with a teacher into students' mathematical thinking is new to many administrators and needs their undivided attention here.
- Student thinking is at the core of a mathematics class in which effective teaching is taking place. Without an understanding of how students are thinking about an idea, a teacher cannot productively plan what pedagogical moves to make. In addition, this kind of reflection on student understandings provides a vital source of generative learning for teachers.

Show Part Two, which contains all three videotaped excerpts (about 2 minutes each), and invite participants to make notes on their transcripts as they wish.

20 minutes

Sharing and Writing about Questions That Invite Collaborative Inquiry Call everyone back together, and ask participants to share an example of student thinking that they would like to understand better and the question they would ask to engage in collaborative inquiry about that student's thinking with Mr. Mamer. Keep track of their examples of student thinking on one sheet of chart paper and the questions they would ask on another sheet of chart paper.

After several groups have had a turn sharing their interests in student thinking and the questions they would ask, ask participants to take 5 minutes to respond in writing to the following:

> *What are the characteristics of questions that are likely to prompt generative learning for a teacher?*
>
> *What does shared inquiry into student thinking make possible for a teacher? For you? For the two of you together?*

Mention to participants that although they have talked about this idea in earlier sessions, this is an opportunity for them to consolidate the thinking they have been doing throughout the course about this important issue.

Ask participants to share their thoughts with a partner.

Let participants know that in the final session of the course, they will be thinking more broadly about an issue they considered briefly in Session 1: the purpose of teacher supervision in their schools and how distributed supervision might contribute to generative learning across roles in their schools. A backdrop for this focus will be the work they have done making sense of specific classroom episodes and engaging in collaborative inquiry with teachers about them.

Overview

For this viewing, you will be showing three short excerpts from the *Designing Packages* classroom episode. Participants will address a set of discussion questions (Handout 36) as they work with the transcripts (Reading 8). The three short excerpts are contained in Part Two of DVD #3.

You will ask participants to form small groups of three or four. The whole group will view the three video excerpts (about 2 minutes each), and then each small group will work closely with the transcript of one of the three excerpts. You should assign the excerpts to groups before the videotape viewing so that groups will know which excerpt will be theirs to analyze as they view it.

First excerpt

In the first excerpt, Ginny is uncertain about how to know whether they have found all of the rectangular prisms that can be made to hold 24 cubes. She has listed several but does not know whether there are more. Mr. Mamer asks Osam what he was doing. Osam explains that he was using the multiplication chart to see what would "times up to 24." Mr. Mamer comes back to Ginny with the question she initially posed about whether her group has come up with all possible combinations. Ginny then wonders whether they should see whether other numbers when multiplied together equal 24. Mr. Mamer leaves them with the suggestion that they work on this.

Osam and Ginny seem to be on the verge of understanding that the dimensions of the rectangular prisms are factors of 24 and that they could use that information to figure out whether they identified all of the rectangular prisms.

Mr. Mamer's instruction is very subtle here. Most of what he says is in the form of questions. He brings Osam into the discussion possibly as a way of pushing Ginny's thinking, although we cannot know this for sure from this bit of footage. When Mr. Mamer sees that Osam and Ginny have hold of the idea of multiplication of factors, he suggests that they work on that idea some more. He leaves them with an open-ended exploration to pursue.

It would be interesting to know how the other two students in this group were thinking about this issue. It would also be interesting to know whether either Ginny or Osam had a firm hold on the idea that working out all of the multiples would give them all of the prisms when they talked about multiplication or whether this idea would become more firmly planted as they continued their explorations.

Second excerpt

In the second section, Libby is explaining that rectangular prisms with identical dimensions would have different surface areas if they were oriented in different directions. She seems to think that if the height, depth, and width were labeled differently, the surface area would be different. Ginny thinks that Libby is wrong and that if the length and width were switched around, the surface area would be the same because it is the same shape, just turned. Libby remembers the cereal box shown at the beginning of the class and concedes that she is wrong—that it was the same box in different orientations. Mr. Mamer picks up the box and illustrates this point.

Mr. Mamer supports Libby by accepting her answer as plausible, and then he calls on Ginny. We do not know why he calls on Ginny in particular. Does he know that she had this worked out already? Does he think she might have worked this out already because of how she usually thinks about mathematics? Given the degree of attention he has been paying to the students' mathematical thinking as they worked in small groups, he might have called on Ginny because he knew that she had already worked this out. But he might have called on her for some reason unconnected to mathematical ideas (i.e., she had not been called on for a while or he wanted to call on someone at her table in order to keep them engaged). In any event, Ginny knows that the surface area would stay the same and is comfortable saying that Libby is wrong. After Ginny gives her explanation, Mr. Mamer does not say that she is right. As he does throughout this footage, Mr. Mamer does not tell students that their conjectures are correct or not; rather, he often repeats what they said, usually in a neutral tone of voice, or asks another question. In this case, he asks Libby what she thinks about Ginny's idea, giving Libby the opportunity to think it through for herself.

Once again Mr. Mamer's contributions to the discussion are mostly in the form of questions, and for the most part his questions invite students to explain their thinking further. For example, he asks Libby, "Why is it that each of the groups are having the same surface areas in their charts?" When Libby responds that if the boxes were oriented in a different way then the surface areas would be different, Mr. Mamer asks, "What would have been totally different?" After he asks Ginny what she thought, and Ginny says she disagrees with Libby, Mr. Mamer asks, "Why is that?" Through these questions, Mr. Mamer is probing the thinking of these two students to find out more. Only at the end of this excerpt, after Libby indicates that she understands that the surface areas of differently oriented boxes with the same dimensions are the same, does Mr. Mamer stop asking questions and pick up the cereal box to make the point that the different orientations do not affect surface area.

It would be interesting to know why Mr. Mamer called on Ginny. It would also be interesting to know even more about how and what Libby was thinking when she thought the surface areas would be different and what Mr. Mamer was thinking about Libby's thinking.

Third excerpt

Grace begins by saying that as the shape of the rectangular prism becomes more cubelike, the height gets taller (see Figure 3). Mr. Mamer asks whether anyone has an idea of why this is so. Libby tries to explain why, but her explanation evolves into a discussion of why the surface area of a rectangular prism gets smaller as the shape of the prism becomes more cubelike. Libby explains this haltingly, and then Evan explains it a bit more clearly: that in the long, rectangular shape, each cube has more faces exposed, but in the shape that is more cubelike, more of the faces are interior. This is clearly a new idea to Mr. Mamer, and he is very interested in understanding it and repeating it so that the rest of the class has the benefit of Libby and Evan's thinking.

It is not clear what Grace means when she says that as a rectangular prism becomes more cubelike the height gets taller and taller, although we can guess at what she is thinking by examining Figure 3 and noting how the dimension labeled "height" goes from 1 to 2 as the box gets more cubelike.

Unlike his more probing approach to working with his students' ideas in the second excerpt, Mr. Mamer does not ask the questions in this excerpt that would give him a fuller picture of either Grace or Libby's thinking. He asks whether anyone can explain why the height increases as the box gets more cubelike, but when Libby shifts the discussion to why the surface area decreases, he does not go back to Grace's idea.

We do not have enough information to know why Mr. Mamer made these teaching decisions—whether he intended to go back to Grace's idea later, whether he thought that pursuing it would take them away from the main mathematical ideas that he wanted them to address, or some other reason. He focuses on Libby and Evan's idea about the cause of reduced surface area. As both Libby and Evan explained parts of their idea,

Mr. Mamer repeated part of what they said. They, in turn, would confirm that they agreed with what he had said. It is almost as if Mr. Mamer were trying to make sure that he understood what they were trying to say and, in making the idea clear to himself, he made it clear for the class as well.

Teachers are always juggling alternatives such as those described above. At any one time there are a number of pedagogical moves that might make sense. As participants think about the complex mathematical thinking that is taking place in this class, they can begin to appreciate the tensions and dilemmas encountered by teachers who, like Mr. Mamer, are working hard to help their students develop their mathematical thinking.

Questions for Collaborative Inquiry with Mr. Mamer

By now, in addition to their consideration of *what* might be fruitful to talk about with teachers, participants have had a range of experiences with *how* to talk to teachers about the mathematics learning and teaching taking place in teachers' classrooms.

In Session 5, participants were introduced to the idea of generative learning and they analyzed their communication modes and showed them in pie charts. They used a collaborative inquiry approach with teachers in a post-observation conference in the role plays they did during the session.

In Session 6, participants observed a videotaped conference between a teacher and a math supervisor who was modeling how to support a teacher's generative growth by using a collaborative inquiry style of talking to the teacher.

For homework for this session, participants conducted a post-observation conference with a teacher in their schools and gained firsthand experience using collaborative inquiry as one of the ways of talking to the teacher.

Finally, after observing the videotaped classroom episode and attending to students' mathematical thinking and the ways the teacher works with their ideas, participants will consider the questions about students' mathematical thinking they would like to explore with the teacher, keeping the idea of the teacher's generative learning in mind.

Up to this point in this session, participants have been attending to *how* Mr. Mamer works with students' ideas. It is likely that they will have questions and comments about the pedagogical decisions he made. For this part of the activity, however, it is important for you to encourage them to focus on questions that target students' mathematical thinking only. It will be important for you to mention this shift in focus clearly at the beginning of this part of the activity and to explain that such a focus has been chosen because student thinking is at the core of a mathematics class in which effective teaching is taking place. Without an understanding of how students are thinking about an idea, a teacher cannot productively plan what pedagogical moves to make. This kind of reflection on student understandings provides a vital source of generative learning for teachers.

On Handout 36, participants are asked to choose one or more examples of students' thinking from the sections of the transcript (Reading 9) about which they have a genuine curiosity and then to form questions to ask Mr. Mamer in the context of a post-observation conference. For example, from the second section of the transcript, they may want to know more about what Libby had been thinking when she said that if the rectangles "sat" in a different direction, the surface area would have been totally different.

Some questions about Libby's thinking that might be productive to explore with Mr. Mamer follow:

> I wonder what Libby meant when she said that if the rectangles were oriented in a

> different way the surface area would have been entirely different.
>
> I wonder whether she thinks that if numbers are multiplied in a different order, the answer is different.

It would also be interesting to know more about Grace's idea from the third section of the transcript. Grace noticed that as the rectangular prism became more like a cube, the height increased. Grace is focusing on what happens to the dimensions of the prism as it approaches a cube, not on what happens to its surface area. We can see that as the rectangular prism becomes more like a cube, one dimension becomes larger, one becomes smaller, and one becomes larger and then smaller (see Figure 3). Whether or not the "height" increases, as Grace asserts, depends on which dimension has been labeled "height."

> I wonder what Grace was noticing that led her to say that the height would increase.
>
> I wonder what she thinks the attributes of a cube are.
>
> What does she seem to understand, and what does she seem to be puzzling about?

At the end of this activity, after participants have shared their questions, they will respond in writing to the following questions and share their thoughts with a partner.

- **What are the characteristics of questions that are likely to prompt generative learning for a teacher?**
- **What does shared inquiry into student thinking make possible for the teacher? For you? For the two of you together?**

Questions that promote genuine inquiry on the part of the teacher tend to be open-ended questions that are firmly rooted in the details of a student's mathematical thinking. They do not have answers that are already known; rather, they invite the generation of hypotheses about the student's mathematical thinking.

A teacher has much to gain by having another set of eyes in the classroom that are focused on students' mathematical ideas and ways to best support their development. When administrators help teachers make sense of students' mathematical ideas, the increased understanding of students' thinking that results can lead teachers to work more productively with students' ideas.

In sharing genuine curiosity about a student's thinking about a particular mathematical idea, an administrator can set a cultural norm in which students' mathematical ideas are considered interesting and important to attend to. In this intellectual culture, close examination of students' thinking can also help the teacher and the observer uncover sometimes subtle aspects of mathematics ideas as part of extending their own learning.

It will be interesting for you to collect participants' thoughts about this and to see to what extent they are appreciating the value of probing into students' mathematical thinking. Have them also consider how this can be a generative process for administrators and teachers alike.

Although participants have had the opportunity to address a similar question in an earlier session, the writing they do about it here will help consolidate the thinking they have been doing through the course about this important issue.

CLOSING
10 minutes

Materials
Handout 38
Reading 10

Homework

5 minutes | **Assigning Homework** For homework, participants will read Sergiovanni and Starratt's *Sources of Authority for Supervision* (Reading 10) and will write about the questions on Handout 38, Homework for Session 8.

Bridging to Practice

5 minutes | **Reflective Writing** Remind participants that Bridging to Practice is designed to allow them to focus on the ideas in the course that seem particularly interesting for them and for their school community. Point out that it provides a framework for planning what they can do in order to move themselves and their schools along with these ideas.

Post the following questions, and invite participants to respond to them:

- *Choose an idea that came up today that you found particularly interesting. What is your current thinking about this idea?*
- *Where is your school now with regard to this idea?*
- *What are one or two things that you, as instructional leader, will pursue to move yourself and/or your school along with this idea?*

Your Own Journal Writing

Within one or two days of teaching this class, after you have had the chance to unwind from the class and perhaps talk with a colleague or friend about how it went, set aside some time to write your own journal entry about the class. We encourage you to write about the ideas held by the participants in your group rather than your own actions and sense of how things worked out. In our experience, writing about what the participants are thinking about will be much more useful to you as you prepare for the next session. Some facilitators use the time when administrators are writing in their own journals to make preliminary notes for this journal writing.

The *Designing Packages* Problem

The students in the classroom episode will be working on the problem below. Take about half an hour to explore this activity. Do as much as interests you. Use graph paper or unifix cubes, as you choose.

You work for the ABC Toy Company, and your job is to find all of the boxes that could be used to ship 24 cubes. (The boxes should fit the set of cubes exactly—no empty space.)

Make a chart that shows the length, width, and height for each box with a volume of 24. Make a sketch of each of these boxes, and determine its surface area.

How many different boxes are there? How do you know that you have identified them all?

Which shaped box uses the least amount of packaging material? Why?

Write a formula for calculating the surface area of a box.

Working with the *Designing Packages* Transcript

In small groups, discuss the following questions about the excerpt of the classroom episode that you were assigned.

1. What mathematical ideas seem to be interesting or confusing to students?

2. What does Mr. Mamer do to help students think through these ideas?

3. What are one or two examples of students' thinking that you would like to understand better?

4. Using these examples, what are some questions that you might like to investigate with Mr. Mamer?

Learning & Pedagogy Observation Guide

Students		
Focus Question	Conjectures	Evidence from Classroom
What kinds of mathematical sense making are students doing?		
• What are the partially formed ideas that students are working with?		
• What ideas are the students particularly engaged in thinking about?		
• What ideas seem to be well understood?		
• What ideas seem to be poorly understood?		
What mathematical ideas seem to be confusing to students?		
• What might the root of the confusion be?		
• What new thinking does the confusion lead to?		
How are students developing their mathematical ideas over time?		
• In what instances do students refer to prior discussion, explorations, or findings?		
• Do students say anything about what they now understand or what they still need to learn?		
• What changes in thinking do you see that the children are not directly acknowledging?		

Learning & Pedagogy Observation Guide

Teachers		
Evidence from Classroom	Conjectures	Focus Question
		How does the teacher work with the sense the children are making?
		• What opportunities does the teacher give students to share their thinking?
		• What does the teacher do to build on their ideas?
		• Does the teacher choose not to pursue certain ideas?
		How does the teacher work productively with students' confusions?
		• What does the teacher do to help students think through confusing ideas?
		How does the teacher show that he or she understands that students' ideas develop over time?
		• What does the teacher do to help students make connections to prior mathematical ideas?
		• How does the teacher work with students who are in different places with respect to any particular idea?
		• What expectations does the teacher have regarding students' levels of understanding based on the way the session ends?
		How does the teacher adjust her teaching based on the ideas that she hears from students?
		How does the teacher attend to all students?

Homework for Session 8

Read *Sources of Authority for Supervision* by Sergiovanni and Starratt, (Reading 10) and think about the following question:

1. How do the ideas in this reading relate to what we have been re-examining in this course, particularly with respect to the two themes of the course, *developing a new eye for what is important to look for in mathematics classrooms* and *rethinking the ways we talk with teachers about what we observe in their classrooms?*

Then write about the following question:

2. What supervisory practices are currently in place in your school? Use Sergiovanni and Starratt's description of sources of authority as a framework for examining the kinds of authority you call upon to support these practices.

Finally, reread what you wrote in Session 1 of this course about what you need to know in order to be a good observer and what you gain from conducting classroom observations. Amplify and extend your response in order to reflect your current understandings.

SESSION
8

Bringing It All Together: Distributing Supervisory Practices and Fostering Generative Growth

Increasingly, there is recognition that generative learning is nurtured when teachers and administrators work together in learning communities, rather than working in isolation. Being part of a learning community allows for the exchange of understanding and perspectives that is central in promoting growth—what Brian Lord has termed "critical colleagueship."[1] When learning is embedded in the entire school community, school becomes a place where not only children, but also adults in various teaching and administrative roles are learning as they do their work. This combination of activity and reflection forms the foundation for generative learning.

This session explores redistributing the responsibility for classroom observation and supervision across different roles in the school. In this way, both the responsibility and the learning that comes with conducting and discussing classroom observations are not limited to those in formal supervisory roles, but are shared by teachers, staff developers, and administrators alike. In this session, participants explore several important related ideas:

1. They rethink the sources of their own authority and capacity for influence.
2. They reflect on how *generative learning* might guide their supervisory work with teachers.
3. They consider how classroom observation and supervision can become a rich context for professional development for both teachers and themselves.
4. They explore how the benefits afforded to both the teacher being observed and to the observer might be distributed across roles in the school.

This session is intended to leave participants thinking about how they might continue to develop their "new eye" as well as use it to support different practices of observation and supervision as central parts of communities of learning.

---

[1]Lord, B. (1994). Teachers' Professional Development: Critical Colleagueship and the Role of Professional Communities. In N. Cobb (Ed.), *The Future of Education: Perspectives on National Standards in America* (pp. 175–204). New York: College Entrance Examination Board.

Overview for Session 8

OPENING page 274 10 minutes	**INTRODUCTION** This is the final session of the course. The Introduction offers a time to summarize what participants have covered in the course and to present the possibilities for professional development and growth that classroom observation and supervision might offer.
ACTIVITY 1 Homework Discussion page 276 20 minutes	**WHAT'S IN IT FOR US? REVISITED** Participants share their new responses to the question, "What do you need to know to be a good observer?" that they wrote about in the first session. Participants consider what they have learned in the course about how to be a good observer in a standards-based mathematics classroom and begin to look at ways their work can be enhanced if they share the responsibility for classroom observations with other professionals in the school community.
ACTIVITY 2 Discussion of Reading page 278 55 minutes	**SOURCES OF AUTHORITY, SOURCES OF INFLUENCE** Participants discuss the Sergiovanni and Starratt article they read for homework. Their discussion focuses particularly on reconsidering sources of their own authority and capacity for influence with the teachers in their schools. They examine the Observation Guide with an eye for how it helps develop different kinds of authority for supervision and how these different kinds of authority can support generative learning.
ACTIVITY 3 Discussion of Scenarios page 284 75 minutes	**NEW IMAGES OF LEARNING FROM CLASSROOM OBSERVATIONS** Participants examine two brief descriptions of alternative ways of using classroom observation and teacher supervision—an adaptation of a "Japanese Lesson Study" and a joint Staff Developer/Principal Observation—to explore how the practice of classroom observation and supervision can open up opportunities for sustained learning when it goes beyond the familiar dyad of teacher/supervisor.
CLOSING page 291 20 minutes	**BRIDGING TO PRACTICE** In the last reflective writing of the course, participants consider how the benefits afforded to both the teacher being observed and to the observer might be distributed across roles in the school.

BIG IDEAS

Participants explore:

- the link between the quality of supervisors' influence and the sources of authority they draw on;
- the role of supervision in supporting both teachers' and supervisors' generative learning;
- the benefit of distributing supervisory practices across roles in the school in terms of extending the capacity for learning among adults in the school

MATERIALS

- ☐ nametags
- ☐ flip charts or overheads
- ☐ markers
- ☐ agenda
- ☐ Participant writing from Activity 1 in Session 1
- ☐ Handout 38, Homework for Session 8
- ☐ Reading 10, *Sources of Authority for Supervision,* by Sergiovanni and Starratt
- ☐ Handout 39, Sources of Authority in the Observation Guide
- ☐ Handout 40, Distributing Authority Across Roles
- ☐ Handout 41, Generative Learning from Collaborative Inquiry
- ☐ Handout 42, Overall Observation Guide

PREPARATION

- Prepare an agenda for the session on an overhead or on flip chart paper.
- Write the prompts for Bridging to Practice on an overhead or on flip chart paper.
- Reflect on how participants' ideas have emerged and developed over the trajectory of this course. What have been important junctures in your discussions as a group? Which issues and ideas have remained constant and which have evolved and moved? Reviewing participants' journal writings as well as your own journal entries from past sessions will ground your analysis of how ideas have developed among the participants in your course.
- Read over participants' writings for Activity 1 if you collected them. A good preparation would be to select a few of the initial responses and think about how you would revise them if you had written them. Think also about how the responses might mean something different to someone who took the course than to someone who didn't take it.
- Read Sergiovanni and Starratt's *Sources of Authority for Supervision* (Reading 10), and consider the discussion questions for Activity 2. Review Nolan and Francis's article, *Changing Perspectives in Curriculum and Instruction* (Reading 1), discussed in Session 1, as it provides important additional perspectives for today's discussion. Read over the Overall Observation Guide and consider what kind of authority it cultivates and how an administrator might use that authority to support generative learning.
- Read the two scenarios for Activity 3 and consider how each builds on the practice of classroom observation but goes further to involve a range of participants who bring their own authority and expertise to the experience.

OPENING
10 minutes

> ***Materials***
> Name tags
> Agendas

Introduction

Before starting the session, post the agenda in a place where everyone can see it.

10 minutes

Introducing the Session Remind participants that this is the last session of the course. Explain again that the course has been structured to enable them to examine two key thematic strands related to classroom observation and teacher supervision. The first session provided an introduction to these strands and the subsequent sessions developed them further. They are:

- reorienting their focus in the classroom in order to attend to the critical aspects of a standards-based classroom: the mathematics content of the lesson, the learning that is happening and the teacher's pedagogical moves, and the nature of the intellectual community present in the classroom.
- rethinking their talk with teachers about mathematics, learning, and teaching during pre- and post-observation conferences and supporting teachers' generative learning by assuming a stance of collaborative inquiry.

Explain that this session brings closure to the course by opening up some possibilities for how they can further use classroom observation and supervision as a context for professional development for both themselves and teachers. In this session they:

- consider the nature of the "authority" that they bring to classroom observation and supervision.
- examine how the responsibility for classroom observation and supervision can be shared across roles in the school.

Mention that this session does not include a videotaped clip. Rather, participants will have opportunities to "digest" several key ideas brought up in the Sergiovanni and Starratt article they read for homework, to consider several additional ones, and to explore the meaning of these ideas for the supervision of mathematics in their schools.

You may also want to take this opportunity to remind participants of important junctures in your discussions as a group, and how issues and ideas have evolved over the course.

Overview

As you reach this last session of the course, you will know your group very well. You should be able to think about where each of the participants is with regard to the key ideas of the course. This session shifts from the very specific focus on observing in classrooms and talking with teachers to looking at the broader role that classroom observation and supervision play in the overall professional development of teachers and administrators. These **Facilitator Notes** help you, as facilitator, make this shift in focus and connect the ideas in this session with the work from the previous sessions.

Remember that the discussion questions that accompany activities in the curriculum are meant as starters for discussions. These notes are meant to prepare you so that you will be able to identify these ideas more easily when they occur in seminar discussion or direct the discussion to them if they do not.

As you introduce this session, you may want to take the opportunity to remind participants of important junctures in your discussions as a group, and how issues and ideas have evolved over the course.

ACTIVITY 1
Homework Discussion
20 minutes

Materials
Handout 38
Journal Writing from Activity 1

What's In It For Us? Revisited

10 minutes

Sharing Responses Ask participants to take out their original writings from Session 1 and their response to question 3 in the homework and to share them with a colleague. Question 3 asked participants to reread what they wrote in Session 1 of this course about what one needs to know in order to be a good observer and what they gain from conducting classroom observations. The original questions are:

In sharing their writings with a colleague, participants can think more carefully about why they revised their responses as they did and can benefit from the reflections of others.

What do you need to know in order to be a good observer?

What do you as an administrator gain from doing classroom observations?

They were to have revised their responses and extended them in order to reflect how their thinking about classroom observation has changed since the beginning of the course.

10 minutes
Whole group

Revisiting Classroom Observations Bring participants together for a short whole group discussion. You can open the discussion by asking someone who has changed his or her responses to share the original responses and the revised ones. Ask whether others revised their responses in similar ways. Use the discussion to explore the different ways participants changed their answers and the insights they have gained throughout the course to date.

You might also ask whether there are any participants who did not change their original responses. If there are, ask them why they did not change or whether what they wrote originally has a different meaning to them now.

Tell participants that in the next activity, they are going to examine the sources of authority for supervision. They will also examine how doing classroom observations can cultivate different types of authority.

Facilitator's Notes
Activity 1

Overview

In the first session, participants responded to two questions as a way of getting them thinking about the role of classroom observation and teacher supervision in their work. These were:

What do you need to know in order to be a good observer?

What do you as an administrator gain from doing classroom observations?

Part of the homework assignment for this session was for participants to revisit their responses to these questions and revise them in light of what they have been learning in the course. Sharing their revisions with someone else in the group provides them with an opportunity to think more fully about why they revised their responses as they did.

The short discussion that follows is intended to promote further reflection on participants' thinking, in particular to trace the evolution of their ideas over the course. By asking whether someone kept the response the same but thinks about it differently, you acknowledge the possibility that some people might have deepened their understanding of what is involved in observing in classrooms. For instance, someone may have written that an effective observer is one who listens well. Now, that participant may understand "listening well" to include listening carefully to the mathematical ideas of the lesson.

As was noted in the **Facilitator Notes** for Session 1, the Observation Guides that have been used throughout this course have focused attention on the mathematical ideas of the lesson, the ways students understand and work with the ideas and the teacher's pedagogical moves, and the nature of the intellectual community of the classroom. As participants revisit what they wrote initially, many of them may revise their responses to reflect these facets of mathematics learning that the Guide has helped them notice.

The **Facilitator Notes** for Session 1 also describe the benefits of classroom observations for administrators when they focus on students' mathematical thinking and on ways teachers support the development of students' mathematical understandings. Some of the benefits include the following:

- They can deepen their understanding of the dimensions of mathematics, learning, and teaching in standards-based classrooms.
- They can bring increasingly focused lenses to their assessment of mathematics education across the school.
- They can consider in increasingly grounded ways what approaches to supervision and to professional development might best support teacher learning.
- They can explore what data from classroom observations are fruitful to discuss with teachers, and what approaches can best support teachers' generative learning.
- These are all components of administrators' own generative learning.

In addition to the above points, participants may mention the enhanced credibility for leadership that comes with understanding about learning and teaching that can be gained from spending time in classrooms. Participants may also comment that their supervisory practice is a context in which they can directly exert influence on classroom practice through what they choose to focus on when they observe in classrooms and talk to teachers in post-observation conferences.

ACTIVITY 2
Discussion of Reading
55 minutes

Materials
Handouts 39 & 42
Reading 10

Sources of Authority, Sources of Influence

20 minutes
Whole group

Examining Sources of Authority Invite participants to share their thoughts about key or significant ideas in this chapter. You might ask:

What stood out for you as you read this chapter?

Next ask participants to review two of the sources of authority that Sergiovanni and Starratt describe: *professional authority* and *moral authority*. You might ask:

How do Sergiovanni and Starratt describe professional authority, the assumptions that underlie it, and the supervisory practice that is based on it?

How do they describe moral authority, the assumptions that underlie it, and the supervisory practice that is based on it?

The question about what stood out for participants gives them the opportunity to share their initial thoughts. It also gives you a sense of what ideas are particularly resonant for participants. This will be helpful information as you guide the group in considering the possibilities of distributing supervisory practices across the school.

On a flip chart, record participants' comments.
Flesh out and extend these comments in order to get to the foundational ideas. Post these comments on the walls to serve as references throughout the session.

If participants comment on the challenges of grounding classroom observation and supervision on these sources of authority or if they identify barriers to using these sources of authority for supervision, you might star them for further consideration in the next discussion.

15 minutes
Discussion in Pairs

Reflecting on Sources of Authority in the Observation Guide Pass out Handout 39, Sources of Authority in the Observation Guide, and Handout 42, the Overall Observation Guide. Ask participants to get into pairs and to look over the Guide carefully and consider the following three questions:

What different sources of authority does the Guide help develop?

How might developing these sources of authority change your relationship with teachers?

In what ways can these sources of authority be used to support generative learning?

20 minutes
Whole group

Classroom Observation as a Source of Authority for Supervision Have participants discuss how the work they have been doing to cultivate an eye for standards-based mathematics classrooms has deepened their professional and moral authority. Use the questions above to guide the discussion.

Use this time to return to the issues and challenges that they see in grounding supervision on professional or moral authority.

Explain to participants that in the rest of this session they will explore classroom observation and supervision as a context for professional development, particularly when the responsibility for it is distributed among different people in the school.

Highlights from the Sergiovanni and Starratt Article

Sergiovanni and Starratt define *authority* as the power to influence thought and behavior. They maintain that the change that is needed in schools requires much more than tinkering with school culture as it currently exists. It requires that basic values of schooling, such as goals, beliefs, and working arrangements, be altered as well as values related to the distribution of power and authority. It is the distribution of power and authority in the context of teacher supervision that they examine in depth in this article.

Sergiovanni and Starratt assert that the present structure and practice of teacher supervision is based on a particular pattern of authority. They argue that changing the current structure and practice of supervision, which they insist is so urgently needed, requires that the authority base for supervision be changed as well. When teaching is viewed as a technical field with teaching practices based on set routines, the role of supervision is seen as providing direction and monitoring for teachers. This approach assumes that sources of authority are external, process-oriented, and management-based. When teaching is viewed as a profession where practices are based on a combination of research, the wisdom of experience, careful analysis of the situation at hand, and a commitment to professionalism, the source of authority shifts to one that is internal, knowledge-oriented, and norm-based.

Sergiovanni and Starratt go on to describe five broad sources of authority:

Bureaucratic authority, "in the form of legal and organizational mandates, rules, regulations, job descriptions, and expectations. When supervisory policies and practices are based on bureaucratic authority, teachers are expected to respond appropriately or face the consequences" (p. 134). When supervision is dictated by bureaucratic authority, teacher effectiveness is measured on predetermined standards, and expertise resides in those higher in the hierarchy. The focus is on compliance, with external monitoring ensuring this compliance.

Personal authority, "in the form of interpersonal leadership, motivational technology, and human relations skills. When supervisory policies and practices are based on personal authority, teachers are expected to respond to the supervisor's personality, to the pleasant environment provided, and to incentives for positive behavior. Personal authority is enhanced by learning how to apply insights from psychology and human and organizational behavior" (p. 134). When supervision is governed by personal authority, success tends to be seen as depending on the supervisor's personality and style, rather than on the intrinsic worth of the ideas.

Technical-rational authority, "in the form of evidence derived from logic and scientific research in education. When supervisory policies and practices are based on the authority of technical rationality, teachers are expected to respond according to what is considered the truth. Research, for example, tells teachers what to do rather than informs the decisions they make about what to do" (p. 134). When technical-rational authority governs supervision, best practices and teacher effectiveness are based on evidence from research. Again, expertise is seen as being external and teachers are told what constitutes effective teaching.

Professional authority, "in the form of experience, knowledge of the craft, and personal expertise. When supervisory policies and practices are based on professional authority, teachers are expected to respond to common socialization, accepted tenets of practice, and internalized expertise. Research, in this case, does not tell teachers what to do but informs the decisions that they make about what to do" (p. 134). While effective teaching practices are derived from research, teachers use research findings to inform their teaching. There is a presumption that teachers can make sound judgments based on the specifics of the situations they face.

Moral authority, "in the form of obligations and duties derived from widely shared values, ideas, and ideals. When supervisory policies and practices are based on moral authority, teachers are expected to respond to shared commitments and felt interdependence" (p. 134). Supervisors direct their efforts toward identifying and making explicit shared values and beliefs, which are then transformed into informal norms that govern behavior. The informal norm system of the school community, rather than external controls, is used to enforce professional and community values.

Sergiovanni and Starratt encourage a shift toward what they call Supervision II, "a process and function that is hierarchically independent and role-free" (p. 145), guided by professional and moral authority. Rather than being carried out according to organizational structure and legitimized by credentials, the process of supervision is seen as "a set of ideals and skills that can be translated into processes that can help teachers and help schools function more effectively" (p. 145). This process can be carried out by teachers and others, as well as by those in hierarchically defined supervisory roles. Sergiovanni and Starratt encourage the deinstitutionalizing of supervision and suggest that it be replaced with formalized "informal supervision" (p. 145), incorporating ideas such as collegial supervision, mentoring supervision, cooperative supervision, and informal supervision.

Sergiovanni and Starratt stress the importance of shifting the ways control is expressed in schools by fostering professional socialization, developing purposes and shared values, building natural interdependencies among teachers, and providing "a kind of normative power that encourages people to meet their commitments" (p. 146). In these situations, teachers become self-managing.

Whole Group Discussion of the Sergiovanni and Starratt Article

◆ What stood out for you as you read this chapter? Many participants are likely to find the ideas embedded in Sergiovanni and Starratt's analysis of sources of authority both enlightening and challenging. Sergiovanni and Starratt define authority as *the capacity to influence.* The capacity to influence is central when supervision is viewed as professional development, with the intent of bringing about growth in teachers.

Given this emphasis, two questions to ask participants are: **What supervisory approaches will best support teachers in learning what they need to learn in order to incorporate central elements of standards-based teaching and learning into their classroom practice?** and **What assumptions about sources of authority are these based on?** Given the complexity of what needs to be learned, it is not enough to inform teachers of a set of steps and procedures they are expected to follow. Rather, in the same way that students need opportunities to construct their own understandings of mathematical ideas, teachers need to be able to construct their own understandings of the significance of new ideas and the relationship of these ideas to their practice. It makes sense, then, that work with teachers be consistent with the tenets of standards-based classrooms and that the work of supervision be grounded in efforts to support teachers in constructing understandings about mathematics, learning, and teaching with their students in their classrooms.

While some of the premises of bureaucratic, personal, and technical-rational authority are easy enough to dismiss, others may continue to be active in participants' outlook on ways in which they, as supervisors, can best support teachers to grow and change.

For example, Sergiovanni and Starratt address the generally held assumption about the need for external monitoring, an assumption that many of your participants might share. They may speak about the importance of providing motivation for teachers to change. They might note that the formalized process of supervision, with its potential for resulting in a negative evaluation, is a feature they rely on to "make teachers listen" to the changes that are being undertaken in the schools. Moving to a perspective in which motivation for change is seen as coming from a combination of internal and community-based imperatives requires a substantial shift in perspectives, in terms of both what is to be learned, how that learning can take place, and what the driving force is in bringing about lasting change.

Sergiovanni and Starratt also note the extent to which many supervisors rely on personal authority to exert influence on teachers, incorporating many aspects of interpersonal skills and motivational techniques. Again, these are characteristics of leadership that participants are likely to believe are central in their practice. Indeed, the authors assert the value of this aspect of supervision of leadership. But they note that it should not hold as prominent a position as it currently does.

Like bureaucratic and technical-rational sources of authority, personal authority "should do no more than provide support for a supervisory practice that relies on professional and moral authority." (p. 137) What participants may find challenging is rethinking the extent to which they are relying on personal authority to motivate change as opposed to seeing it as an important component of a process that is primarily professionally and morally driven.

◆ **How do Sergiovanni and Starratt describe professional and moral authority, the assumptions that underlie them, and the supervisory practices that are based on them?** Brief descriptions of these forms of authority are included above in the **Highlights** section.

Examining the Guide for Sources of Authority and Support of Generative Learning

Participants may view the Observation Guide primarily as a tool for directing attention to central features of standards-based classrooms. They may not yet have appreciated how the Guide can serve as a professional development tool for *themselves.* As they think about the sources of authority and recognize how central moral and professional authority are to their work as instructional leaders, they can look at the Guide in a new light.

This part of the activity helps tie the work they have been doing to cultivate a new eye for standards-based classrooms to their broader roles in their schools as instructional leaders. It also helps participants gain a deeper appreciation of the importance of being learners while also being leaders.

◆ **What different sources of authority does the Guide help to develop?** The Guide is designed to help those who use it to think more deeply about the central features of standards-based classrooms. It helps participants develop their observing and inquiry skills about the mathematical thinking and learning that are taking place in the classroom. In that sense, they are developing their professional authority. The Guide also draws attention to the features of an intellectual community, thereby legitimizing the values on which the notion of moral authority rests. Specifically, the Guide draws attention to how the students show respect for one another and how the teacher respects the students' ideas. It also values students and teachers using one another as resources and working together to make sense of mathematical ideas. Thus, as adminstrators use the Guide, they are developing the basis for their own "moral authority" in that they are developing a deep appreciation for the importance of collaborative inquiry. The Guide invites those who use it to engage in collaborative inquiry and encourages shared beliefs and values about its importance in the school community.

◆ **How might developing these sources of authority change your relationship with teachers?** In developing their own professional and moral authority, administrators can change their relationships with teachers. While administrators may still use their own bureaucratic authority to establish and maintain appropriate observation and supervision structures in the school, developing their own professional and moral authority changes their legitimacy vis-à-vis teachers. It allows them to strengthen their commitment to the values and beliefs that underlie standards-based pedagogies and to develop a deeper understanding of what these beliefs and values look like when implemented.

◆ **In what ways can these sources of authority be used to support generative learning?** Generative learning is not something that can be imposed. Being a generative learner, as Franke and colleagues (1998) have described, is something that comes from within the person. Administrators would not be able to call upon school rules, mandates, or regulations to make anyone, including themselves, into a generative learner. Cultivating—and using—professional and moral authority generally entails engaging in collaborative learning, which itself is necessary for generative learning.

Whole Group Discussion

The whole group discussion revisits the questions above. Participants may very well have a range of responses to the questions above, particularly in thinking about the different types of authority that the Observation Guide develops. It might be very useful to focus their discussion on the knowledge that is gained by using the Guide and the changes in teacher/supervisor relationship that is possible when an administrator uses the Guide to reshape how to pay attention in mathematics classrooms.

In this discussion, participants may raise questions about the other sources of authority discussed in the article. It is important to emphasize, as Sergiovani and Starrett do, that all sources of authority have their place. Administrators certainly need to be able to call upon rules, procedures, or state and federal mandates to help establish critical learning structures and processes. They also need to have accurate knowledge of research on teaching and learning as well as be sensitive to the interpersonal needs of their teachers. Some participants, perhaps more used to an adversarial relationship with teachers, might feel that resting supervision on moral or professional authority would be too idealistic. You can explain that this exploration has been focused on the possibilities for professional growth that come when administrators rest their authority as instructional leaders on professional or moral authority. Further, you can point out how the work that they have been doing to cultivate their eye for standards-based classrooms indeed develops their professional and moral authority vis-à-vis teachers. It is important to acknowledge, though, that resting their supervisory practice on professional or moral authority takes time. Administrators not only need to deepen their own understanding of mathematics, learning, and teaching, but they also need to cultivate trust and respect from teachers and other school community members. You can also point out that the next activity will look at ways in which administrators can continue to develop their own professional and moral authority by putting themselves in the position of "learner" within their own schools.

ACTIVITY 3
Discussions of Scenarios
75 minutes

Materials
Handouts 39, 40 & 41

New Images of Learning from Classroom Observations

20 minutes
Small group

Sharing School-based Supervision Practices Ask participants to get into groups of three or four with members from different schools or school systems. Invite them to do the following:

- Share their responses to question 2 on Handout 39 about the supervisory practices that are currently in place in their schools or school districts and the kinds of authority they call up to support these practices.
- Make a list of challenges and issues that would arise when framing supervision in professional and moral authority as a context for professional development in their schools.

The questions about professional and moral authority are intended to help participants connect to the core ideas for each source of authority so they can draw on them in their exploration of issues, dilemmas, and possibilities in their supervisory practices.

For the last 10 minutes of this discussion, ask participants to list the issues and challenges they identified. Record their ideas on flip chart paper to refer back to later in the session.

35 minutes
Small group

Images of Learning from Classroom Observations Tell participants that they will now examine two scenarios of classroom observations that go beyond the more familiar dyad of teacher/supervisor. One is doing joint staff developer/principal observations and the second is doing "Japanese Lesson Study" cycles. Note that while the scenarios they read and discuss are not actual events, both approaches are being experimented in different school districts, including District 2 in New York City.

Have participants stay in their small groups and distribute Handouts 40 and 41. Tell them have have about 15 minutes to read and discuss the scenario in Handout 40. Have them address the following questions:

What do Ms. Cataldo and Mr. Silver each contribute to the joint SD/principal observation?

What do the teachers gain by having Mr. Silver and Ms. Cataldo do joint observations?

How might participating in these joint observations improve Mr. Silver's capacity to do both supervision and formal evaluations?

After about 15 minutes, have the participants look at the second scenario, on Handout 41. Again, have them take five to seven minutes to read the scenario and then spend about ten minutes discussing the following questions:

What can teachers gain by committing time to doing lesson study cycles?

What can Mr. Silver learn by participating in such a lesson study cycle?

What can Mr. Silver gain by supporting the practice of lesson study in his school?

What are some of the challenges of doing lesson study cycles so that teachers can learn from them?

20 minutes
Whole group

Sharing and Discussion Have two flip charts ready, one labeled "Japanese Lesson Study" and one labeled "Joint SD/Principal Observation." Use the questions below to guide a discussion that looks at how these two images of classroom observation and supervision can extend the learning from observing in mathematics classrooms. For the question about how these structures can help address issues that participants identified earlier, refer them to the list you made earlier.

How does authority and expertise get distributed in each of these structures of classroom observations?

What kind of learning is taking place and who is learning what from whom?

How might these structures help to address some of the issues participants raised earlier when they were considering the consequences of adopting supervisory practices that rest on professional and moral authority?

How might administrators' participation in one of these two structures build their credibility and authority vis-à-vis teachers in the school?

Use the flip charts to record the similarities and differences between the two different structures. As participants talk about the two scenarios, you might also try to draw out ways the structures complement each other.

Tell participants that they will now do the final journal writing of the course. Note that more time has been allotted so that they will have time to share some closing reflections.

Introduction

In this activity, participants read and discuss two scenarios that describe a principal's participation in alternative ways of using classroom observation and teacher supervision as contexts for learning. While the two scenarios are not of actual people, the approaches described in them are being experimented with in different school districts, among them District 2 in New York City. You can see the Resource List at the end of the course materials for some additional references on these structures. Both scenarios are set in the same school to project the idea that this school and its district are engaged in a comprehensive professional development program in K–8 mathematics.

These scenarios build on the idea that Nolan and Francis propose in their article (Reading 1) that supervision should become more group-oriented than individual-oriented. They suggest that supervision be viewed as a function to which many people in a school can contribute, through processes such as group supervision, peer coaching, or colleague consultation, to help address content-specific issues. They describe group-oriented supervision in which teachers gain from one anothers' experiences and insights as they learn about the instructional processes and improving student learning. Group and individual goals become interdependent, as supervisors and teachers collaborate and work together to achieve these goals.

The idea that classroom observation and supervision can be fertile contexts for professional development dovetails with Franke *et al.'s* (1998) idea of what supports generative learning. In their article (Reading 7), they note that a focus on children's thinking provides a fertile context for supporting teachers' generative growth and that such a focus can also promote conversations that take place beyond the walls of individual teachers' classrooms.

> We are not proposing that children's thinking is the only avenue for teachers to become generative learners; however, this focus does have characteristics that support generative growth. Children's thinking is available to teachers in their own classrooms daily. There are regularities in the strategies that children describe and principled ideas about these regularities. Additionally, this focus provides teachers with topics to discuss with each other about their classrooms and their students. Teachers can create communities of learning that focus on students' thinking. They can consider how their students are thinking about the mathematics, what this thinking may mean in terms of students' mathematical knowledge, how they can learn more about their students' thinking, and how they might provide opportunities for students to build on their thinking. Because teachers are focused on their students' thinking, the learning communities these teachers create include their classrooms. Consequently, these broadly inclusive learning communities encourage teachers to engage in inquiry focused on children's mathematical thinking with their students, as well as with their colleagues and themselves.[2]

Throughout the course, participants have considered how such a focus on student thinking can also be embedded in the supervisory/classroom observation process, a structure that is already in place within schools. This activity looks at how extending the supervisory process to include a greater range of school people in the practice of classroom observation and teacher supervision can greatly enhance the benefits across the school community.

[2](Franke et al, 1998, p. 107)

Small Group Work

Sharing Supervisory Practices The activity starts by having participants share what they wrote for homework about the supervisory practices currently in place at their school. Participants, especially principals, are likely to talk about their processes for supervising teachers one-on-one and their procedures for evaluating teachers. They may also talk about practices such as peer coaching. As participants share their descriptions, circulate to the different groups to get a sense of the range of practices. This will help you get a sense of how familiar or different the scenarios that they will read will be.

Issues As they share their current practices, participants are also asked to think about using supervision that is grounded in moral and professional authority. They are likely to mention a number of issues that come up as they think about changing supervision to be more driven by these sources of authority. Among others, these might include the following:

- Limited time to engage in this kind of thorough supervision, in terms of how much time can be made available for frequent enough observations and substantive follow-up conferences;
- Some teachers' low motivation for embarking on substantive change. Participants may note the difficulty of getting some teachers on a course of learning without using the threat of poor evaluations;
- Teachers being used to what Sergiovanni and Starratt call the Supervision I "Just tell me what to do" approach. Participants may be concerned with how to get teachers positioned to benefit from the opportunities that such a different approach would afford;
- Building capacity in such a way that teachers become resources to one another and contribute in substantive ways to their colleagues' professional growth;
- Managing the pressures that the system of evaluation places on record-keeping for "making a case," which makes it difficult to move ahead with supervision as a context for professional development.

Discussion of the Staff Developer/Principal Observation Scenario

◆ What do Ms. Cataldo and Mr. Silver each contribute to the joint SD/Principal observation? A staff developer like Ms. Cataldo brings content knowledge and an understanding of how children develop mathematical ideas. She might bring, for instance, knowledge of a particular strand of math at a higher or lower grade level or experience with the particular grade level of the class being observed. She also brings skill in working with teachers to deepen their understanding of mathematics, teaching, and learning. A principal like Mr. Silver brings knowledge of his school and its learning culture. He also might bring a more general capacity for instructional leadership, knowledge of the broader social and political context of the school, and a genuine desire to learn about children's mathematical thinking. While a staff developer such as Ms. Cataldo should certainly bring a curiosity about children's thinking, a principal like Mr. Silver can use what he does not know to open up investigations into mathematical ideas.

◆ What do the teachers gain by having Mr. Silver and Ms. Cataldo do joint observations? Having both a staff developer and a principal involved in the process of classroom observation and teacher supervision offers the teacher a broader range of perspectives on his or her teaching. Each observer can collect data, ensuring that there will be multiple data sources and a range of data for the post-observation conference. This feature links to Nolan and Francis's notion of using a variety of data sources to capture a lesson as it unfolds over time. In addition, having more than one observer can bring a greater variety of experience and expertise to classroom observations.

◆ **How might participating in these joint observations improve Mr. Silver's capacity to do both supervision and formal evaluations?** Participating in a joint SD/Principal observation can help a principal such as Mr. Silver learn how to pay attention to and engage in collaborative inquiry. The staff developer and teacher can offer insights into how content and pedagogy are linked through specific examples from the lesson. The joint observations can also support a principal's capacity to do evaluations by deepening his or her understanding of what to pay attention to when doing a formal observation. For instance, many principals use an evaluation form that asks very general questions. In working with a staff developer, a principal can gain a deeper understanding of how the questions apply in specific content areas. By participating in such joint observations, a principal's evaluations may become less mechanical and more substantive. While he or she still would have to follow the formal procedures established by the district and union, the principal would be more able to give teachers feedback about what they are doing well and about what to focus on as they set their own teaching goals.

Discussion of the "Japanese Lesson Study" Scenario

"Japanese Lesson Study" has become better known in the United States in recent years. It has been described by Stigler and Hiebert in their book, *The Teaching Gap* (1999), and was highlighted in the Third International Mathematics and Science Study (TIMSS). "Japanese Lesson Study" is an involved process that takes time and commitment to establish. The basic structure of a "Japanese Lesson Study" cycle includes four main components:

1. Teachers identify a teaching goal or an instructional issue they want to explore.
2. Teachers plan a study lesson that includes a detailed lesson plan.
3. One teacher teaches the lesson and the others observe the lesson.
4. Teachers meet to debrief the lesson. They analyze what went well and what did not go well and develop ideas for revising the lesson.

◆ **What can teachers gain by committing time to doing lesson study cycles?** In articulating central components of what they call Supervision II, Sergiovanni and Starratt describe a school community that fosters professional socialization, develops shared purposes and values, builds natural interdependencies among teachers, and establishes "a kind of normative power that encourages people to meet their commitments" (p. 146). These components guide the formation of what they call "formalized informal supervision," the notion of distributing supervision for mathematics across roles in the school community.

The structure of "Japanese Lesson Study" provides a formal structure in which teachers learn together and support their own learning as they plan and revise lessons. It is based on the assumption that teachers need to be learners too—they need to be engaged in ongoing investigations of the content of what and how they teach.

While the teacher who actually teaches the study lesson makes herself vulnerable, the responsibility for how the lesson is designed and taught rests on all of the teachers who participate in planning it. This structure allows *teaching itself* to become the focus of investigation rather than any one particular teacher's teaching practice. In this sense, teachers can learn about teaching as a collaborative endeavor and can step outside of the more familiar practice of evaluation.

◆ **What can Mr. Silver learn by participating in such a lesson study cycle?** While it is not reasonable to assume that principals would have time to participate in all of the lesson study cycles that a school might do in a school year, they would benefit by participating in at least one of them. First, principals would gain a better understanding of the process and be more prepared to support it in the wider community. Secondly, they would gain a deeper appreciation of what is entailed in thinking through the ideas of a lesson and teaching it. Third, principals would have a deeper understanding of who their teachers are and how they engage in their work.

◆ **What can Mr. Silver gain by supporting the practice of lesson study in his school?"** Participants may note that administrators have only a limited amount of time for supervision. Establishing a process like "Japanese Lesson Study," while time-intensive, may help to address administrators' limited time for classroom observation by expanding the core of individuals prepared to observe in mathematics classrooms. While the formal lesson study cycle is done as a group, teachers who participate are more likely to be prepared to observe in classrooms on their own. They may also be able to help support other practices such as peer coaching. The process can serve as a way to build a community commitment for enhancing the teaching of mathematics. Having such a process in place may also remove the tension of evaluation from at least some of the supervisor/supervisee relationships.

◆ **What are some of the challenges of doing lesson study cycles so that teachers can learn from them?** Because the idea is new, participants may find that there are many challenges to making a process like "Japanese Lesson Study" a permanent part of their professional development program. These concerns are not unfounded because instituting such a process requires both restructuring and re-culturing schools. A few specific challenges that participants might raise include the following:

- "Japanese Lesson Study" is a new idea in the United States. Many schools lack the structures needed to support it. For instance, teachers need a common planning time and coverage for when they are observing the lesson being taught. This point was deliberately written into the scenario to highlight its importance.
- Teachers may have an underdeveloped sense of what are pedagogically significant or worthwhile problems to investigate. Teachers who are still learning how to think about their practice may not initially be able to develop productive study lessons. Some teachers may focus more on the behavioral aspects of the lesson such as how well they ask questions or whether or not all of the students are engaged.
- Teachers may not realize the importance of developing investigations that link content and pedagogy. For many teachers, investigating teaching practice involves looking only at their own pedagogical practices and not at the content that they teach. The scenario incorporated both pedagogical and content concerns to highlight this point as well.

Whole Group Discussion

This whole group discussion is intended to pull together ideas from this activity and the previous one about the sources of authority, distributed supervisory practices, and generative learning. It is also an opportunity for participants to think about how the work that they have been doing throughout the course to develop their eye for standards-based classrooms and rethink how they talk with teachers fits into the broader context of the school as a learning community.

◆ **How does authority and expertise get distributed in each of these structures of classroom observations?** This question is intended to help participants think about the ways in which each of these two structures—joint SD/principal observations and "Japanese Lesson Study"—contribute to distributing the responsibility for classroom observation and supervision. In both of the scenarios, the principal still retains responsibility for doing formal evaluations but is able to be a "learner" when participating in the other structures. The expertise is shared among teachers, staff developers, and administrators and the key sources of authority are professional authority and moral authority.

◆ **What kind of learning is taking place and who is learning what from whom?** This question is intended to help participants think more carefully about how administrators can be both instructional leaders and *learners* within the context of classroom observation and teacher supervision. In both scenarios, the learning is distributed among the participants. In the joint observation scenario, for example, the teacher draws from the mathematics specialist and the principal to help her think about the mathematical ideas the students are working on. The mathematics specialist and principal learn from the teacher about how individual students are making sense of ideas. The principal is able to learn from the math specialist what the really important aspects of a mathematics classroom are.

In the lesson study scenario, the teachers learn together and support each other's learning about how to plan together, reflect on their students' learning, and rethink their teaching. To the extent that the lesson study structure focuses on students' thinking and supporting students' learning, it has the characteristics that can support generative learning for teachers. When a principal participates in such a cycle, he or she is also drawn to look carefully at the connections between the ideas of the lesson, the pedagogical strategies, and the students' understandings. A principal who makes the time to participate in such a cycle gains a deeper appreciation for what it means to make the school a culture of learning not only for the students but for the teachers and staff as well.

◆ **How might these structures help address some of the issues participants raised earlier when they were considering the consequences of adopting supervisory practices that rest on professional and moral authority?** A lesson study structure in particular can help to address some of the issues participants may have identified in the earlier discussion. For example, if the issue of teachers' low motivation for embarking on substantive change has been raised, participants may recognize how the group nature of lesson study can provide a needed stimulus to such teachers. Such a structure may also help teachers feel an increasing sense of commitment to the values and norms that the group sets. In addition, the teachers who participate in a lesson study group can be resources to each other and in this way, teachers will be building capacity within themselves for contributing to one another's professional growth.

◆ **How might administrators' participation in one of these two structures build their credibility and authority vis-à-vis teachers in the school?** This question asks participants to think about how administrators' willingness to put themselves in positions of learners rather than experts affects their relationship with teachers. This course has continually stressed that administrators can best support teachers when they themselves are learners and engage in the process of collaborative inquiry about students' mathematical ideas.

When administrators are learners in their own schools, they build their own professional authority and credibility with teachers. Teachers are more likely to trust the ideas and views of an administrator who, as a learner, is gaining the understandings that will allow him or her to participate knowledgeably in discussions about learning and teaching that are taking place in their schools. Furthermore, administrators who have formal evaluation responsibility need to have opportunities to observe classrooms and talk with teachers about what and how they teach when the purpose is not evaluation. In that way, they are preparing themselves to be thoughtful and responsive evaluators as well.

CLOSING
20 minutes

Bridging to Practice

10 minutes

Reflective Writing Because this is the last session, this journal reflection is a little longer. It includes a time for participants to share closing reflections with the whole group. Invite participants to write their journal reflections about what they are coming away thinking about from today's session, and from the course as a whole. Remind them that the notion of generative learning implies that the learning is a continuous and continual experience. Point out that rather than thinking they have just learned a new approach to observation and supervision, they should be thinking that they are *just starting to learn* about classroom observation and supervision. Post the following writing guidelines on an overhead or a flip chart:

What idea that you encountered in today's session do you think is particularly salient? What makes this idea interesting or intriguing to you?

Stepping back and thinking about the course as a whole, what ideas about classroom observation and teacher supervision do you come away from this course thinking about?

How might you continue to develop your own observation and supervision skills?

10 minutes

Closing Reflections As a way of closing the course, ask participants to share a few thoughts about what observation and supervision skills they would like to deepen, how they imagine they will continue to deepen them, and how they now view the role of classroom observation and supervision in their work as instructional leaders.

Overview

At the end of this session, you will ask participants to take twenty minutes to write in their journals about the ideas from this session that were particularly salient for them. The purpose of this writing is for participants to have some time to reflect on the ideas discussed in the class and to articulate for themselves something that is important to them about these ideas. It also helps them make the bridge between these ideas and their own work as administrators.

You also will be asking them to step back, think about the course as a whole, and write about the ideas about classroom observation and teacher supervision that they come away from this course thinking about and working on. Participants may have cumulative reflections to put into writing, given that this is the last session of the course. You may have an alternative way of phrasing this question that would tailor it to your particular group.

This last journal writing also leaves a few minutes for participants to share a closing reflection with the group. This sharing gives everyone a chance to hear how others are thinking about continuing to work on the ideas in their own practice.

Your Own Journal Writing

As this is the last session, you might find it interesting to write a closing reflection afterwards. In this, you can note how different participants' ideas about classroom observation and teacher supervision changed or evolved over the course as well as what you learned. You might, for instance, write about how your understanding of what makes a good observer changed as you facilitated the course. Because you might also have been following the thinking of particular participants, you could reflect on their learning specifically and use that to think about your own facilitation. What facilitation issues came up as you worked to push their thinking along? This reflection could be particularly helpful for you were you to facilitate another *Lenses on Learning* course or a similar professional development program.

Sources of Authority in the Observation Guide

Carefully read through the Observation Guide and consider the following questions:

1. What different sources of authority does the Guide help to develop?

2. How might developing these sources of authority change your relationship with teachers?

3. In what ways can these sources of authority be used to support generative learning?

Distributing Authority Across Roles: Doing Joint Staff Developer/Principal Observations and Post-observation Conferences

Myles Silver, the principal of a medium-sized K-5 elementary school in a mid-sized city, had made a commitment to re-evaluate the professional development available to teachers at his school. Improving elementary mathematics education in the district was a high priority that year, and the district had just hired a mathematics specialist, Eva Cataldo.

Ms. Cataldo brought with her many years of experience. She had worked as a classroom teacher in a neighboring district for seven years before she came to this district, where she worked as a math coach in her school. She developed her skills as a math coach through a series of extended professional development opportunities in which she collaborated with other teachers to explore the mathematical ideas in the elementary curriculum and the ways children come to understand these ideas.

As a math coach, she would observe in teachers' classrooms, work with the teachers to understand the key ideas of the lessons they were developing, and help them learn how to build their lessons from where the students were. Now, she was taking on even broader responsibilities as an elementary mathematics specialist for an entire district. Not only would she still observe and coach teachers individually but she would help the district develop a systematic and coherent professional development program in mathematics across its eight elementary schools.

At the beginning of the school year, Ms. Cataldo and Mr. Silver met to review their plans for the year. She said that she planned to do a series of supervisions with each of the teachers in which she would focus on engaging in collaborative inquiry about the mathematical ideas that the students had. She really wanted Mr. Silver to join her in some of these observations and conversations. Initially, he did not find the idea appealing. He already was responsible for doing formal evaluations and believed he did not have time to do joint observations with Ms. Cataldo. However, as a former language arts specialist, he knew how much more he noticed when he observed language arts lessons than when he observed math lessons. He recognized that he would call upon his deep understanding of the content in a language arts class, whereas in a math class he would just follow the teacher's explanation of how to "do" the problem without fully understanding the underlying ideas. He had a hunch that he could benefit from observing with Ms. Caltaldo.

He finally agreed to do three joint observations with Ms. Cataldo. They would observe one class in kindergarten, third grade, and fifth grade. This would give Mr. Silver an opportunity to consider the growth of mathematical ideas over the grades.

Prior to each observation, Mr. Silver, Ms. Cataldo, and the teacher would meet for a pre-observation conference. They would talk about what the teacher was planning to do, what the intent of the lesson was, and what mathematical ideas were at the core of the lesson. They would also talk about what the teacher knew about the children's current understanding and beliefs about the topic as wel, as what might be confusing to students. For example, when they met with the kindergarten teacher, she said she was planning a lesson on patterns. They talked about two central ideas of patterns: that they were predictable and that they were made of repeating units.

Ms. Cataldo and Mr. Silver would then observe the lesson with an eye on the mathematical ideas. They would both take notes during the lesson, focusing on what they found intriguing. Within the next day or two, they would meet with each other and then with the teacher for a post-observation conference. Mr. Silver noticed how Ms. Cataldo never started the conference by talking about what the teacher could have done. Rather, she always listened carefully as the teacher talked about what the students did and what they understood. Further, she tended to open her comments by picking up on the teacher's comments and focusing on what the students actually did and what she thought they were understanding or grappling with.

After doing the three joint observations, Mr. Silver realized how much more attentive he was when observing a mathematics lesson. In the pattern lesson, for instance, Mr. Silver was struck by how differently he listened as one child tried to explain how a tower built by alternating blue and yellow cubes was similar to one built by alternating red and green cubes. He was more aware of how the child was beginning to notice the more general unit common to both towers — one cube of one color followed by one cube of a second color. He also understood better how, as a generalist, he could supervise teachers as well as do more responsive evaluations. In particular, he found out how his own limited math knowledge could actually help create openings in the conferences as he shared his genuine curiosity about what the students were learning. And, in the process, he was learning more mathematics himself.

1. What do Ms. Cataldo and Mr. Silver each contribute to the joint SD/Principal observation?
2. What do the teachers gain by having Mr. Silver and Ms. Cataldo do joint observations?
3. How might participating in these joint observations improve Mr. Silver's capacity both to do supervision and formal evaluations?

Generative Learning from Collaborative Inquiry: Participating in a Lesson Study Cycle

Myles Silver is a principal of a medium-sized K-5 elementary school in a mid-sized city. At a faculty meeting at the beginning of the school year at which they were reviewing the professional development opportunities in their school, several teachers suggested that they might explore doing "Japanese Lesson Study" in their school. They had been reading about them and the new mathematics specialist, Eva Cataldo, had mentioned that a nearby district was starting to do them as well.

Mr. Silver was intrigued with the idea but was not sure how it could work in his school. Would it benefit the students enough to justify devoting the amount of time of the many teachers needed to do it? Yet, he put aside his doubts to hear more about it. Ms. Cataldo described the process of a "Japanese Lesson Study" cycle: Prior to choosing a lesson to develop and teach, the teachers set a goal that they want to achieve with their students The study lesson, the core of the lesson study cycle, is designed to address that goal. It is planned collaboratively by a group of teachers over a period of time, taught in a real classroom by one of the teachers in the group and observed by the other teachers, recorded in some way such as by audiotape or videotape, and discussed afterward. The debriefing session is a critical part of the learning. Those teachers who planned and taught the lesson describe what they did. Those who observed the lesson comment on what seemed to work well and what did not go so well. The group then discusses ideas for revising the lesson.

Mr. Silver agreed that they could pilot the program that year. Because of the time involved in doing each of these cycles, everyone realized that initially they could not do them as a whole school. They decided that the grades would pair off: K/1, 2/3, and 4/5. With three teachers at each grade level, there would be six teachers for each group. Each group would plan two cycles for the year. Ms. Cataldo would join each group for their first cycle and the teachers themselves would be responsible for facilitating their second cycle. The starting times for each grade-level would be staggered to accommodate Ms. Cataldo's limited time.

They would pick a focus and do one cycle and use the post-observation discussion to reflect on how the lesson went and to sketch out directions for the revised lesson. They would then meet several times to develop the revised lesson. The revised lesson would be taught by a different teacher to a different class. This lesson would constitute their second cycle. The two cycles together would happen over a period of about three months. The teachers agreed that they would meet after school for an hour several times to plan the first lesson together. The teachers for each grade level already had common prep time during which they could do some work to prepare for the after-school meetings. Ms. Cataldo would work with Mr. Silver to make sure that the teachers who were observing would have classroom coverage during the time when the study lesson was taught.

As the school year got underway, Mr. Silver began hearing more about "Japanese Lesson Study" from a variety of sources such as newsletters and professional journals. Already part of Japanese school culture, it was an up-and-coming idea in the United States. He decided that he wanted to join one of the grade-level pairs for one of the sessions. Initially, the teachers themselves were not sure they wanted Mr. Silver to join them. This was their own professional development. And they were aware that doing one of these cycles made them very vulnerable, especially if they were going to be the teacher teaching the study lesson. If Mr. Silver joined them, not only would they be exposing their thinking and their teaching to one another but to their principal as well. Ms. Cataldo told the teachers that participating would be very informative for Mr. Silver and reminded them that he would be joining them not as a supervisor but as a learner himself. She assured them she would take responsibility for reminding him of that if he began to act more like a supervisor.

The teachers from the fourth and fifth grade agreed for Mr. Silver to join them. They decided that they wanted to work on the topic of fractions and focus specifically on the teaching of equivalent fractions. They pondered what kind of lessons worked well in helping students grapple with the idea of equivalence and what they needed to know to be prepared to respond to the ideas that students found confusing or puzzling. They also had a number of questions about how students develop a deep understanding of equivalence. What was hard for students to understand about it? How do they come to understand that, for instance, $\frac{3}{4}$ and $\frac{9}{12}$ represent the same amount of the same whole? How can students make sense of what is going on when using the mathematical procedure of multiplying the numerator and the denominator by the same number; for instance, multiplying $\frac{3}{4}$ by $\frac{3}{3}$ to get $\frac{9}{12}$?

In the planning sessions, the teachers first spent some time delving into the mathematics of equivalent fractions themselves. They explored how multiplying the numerator and denominator by the same number kept the value of the quantity the same but increased the segments the whole was divided into. Mr. Silver found these discussions intriguing but he was confused. "Why didn't it work when you multiplied a fraction by $\frac{0}{0}$?" he asked. The teachers had not actually thought of this. In discussing it, they all came to realize that when you multiply the numerator and the denominator by the same number, you are multiplying it by one, but $\frac{0}{0}$ is equal to zero. In fact, as one teacher pointed out, it is not even a mathematically meaningful expression.

The lesson that they designed asked students to develop diagrams that showed the relationships between fractions such as $\frac{1}{4}$ and $\frac{3}{12}$ or $\frac{1}{3}$ and $\frac{6}{18}$. The intent was to have students construct a deeper understanding of equivalence by seeing the equivalence in the diagrams. They also hoped that students would see how both the number of pieces in the whole and the number of pieces in the segment increased by the same proportion. For example, in going from $\frac{1}{4}$ to $\frac{3}{12}$, there would be three times as many pieces in the segment and three times as many pieces in the whole.

It was fortuitous that the teachers had already discussed why multiplying a number by $\frac{0}{0}$ does not result in an equivalent fraction. During the discussion, one student did notice that if you multiplied the top and the bottom of one fraction by the same number, the result was the second fraction in the pair. This led them to examine whether this was the case with all of the fractional pairs. One student then asked if it would work with $\frac{0}{0}$. The teacher used the opportunity to have students explore what it means mathematically to multiply any value by 1 and by 0, and then to compare this to what happens when you *add one* or *add zero* to any number. Yet, it left her and the students still wondering how *multiplying* both the numerator and the denominator by the same number actually kept the quantity the same. Instead of having one circle divided into 8 segments, why didn't they get two circles divided into four segments each when they multiplied $\frac{1}{4}$ by $\frac{2}{2}$?

In the debriefing session, the teachers explored how the lesson helped students develop a better sense of equivalence, but still left them puzzling about how operations with fractions actually work. Students could see from their diagrams, for instance, that $\frac{1}{4}$ of a circle was equivalent to $\frac{2}{8}$ but they still didn't fully grasp how multiplying $\frac{1}{4}$ by $\frac{2}{2}$ really did mean that the quantity stayed the same. They agreed that, in revising the lesson, they would focus more on the nature of operations in fractions, recognizing that drawing the diagrams by themselves would not lend itself necessarily to seeing the connection to multiplying by 1. They also agreed that they would continue exploring the mathematics carefully before the lesson. They came to appreciate more how doing so would help them have a better idea of the range of mathematical issues that could come up in the session, and they would be more prepared to facilitate discussions that are initiated by the students themselves.

Questions

1. What can teachers gain by committing time to doing lesson study cycles?
2. What can Mr. Silver learn by participating in such a lesson study cycle?
3. What can Mr. Silver gain by supporting the practice of lesson study in his school?
4. What are some of the challenges of doing lesson study cycles so that teachers can learn from them?

Observation Guide

Students		
Focus Question	Conjectures	Evidence from Classroom
Math Content		
• What mathematical ideas are embedded in the lesson?		
• What makes this worthwhile mathematics?		
Learning		
• What kinds of mathematics sense-making are students doing?		
• What mathematical ideas seem to be confusing to students?		
• In what ways can you see that the students are developing their mathematical ideas over time?		
Intellectual Community		
• How are students showing respect for one another's ideas?		
• How do students use each other as resources as they make sense of mathematical ideas?		
• What evidence beyond raised hands do you have that students are engaged?		

Observation Guide

Teachers		
Evidence from Classroom	Conjectures	Focus Question
		Knowledge of Content
		• What does the teacher seem to understand about the mathematics?
		• What is the teacher's long-term mathematical agenda?
		• What does the teacher seem to understand about the development of children's ideas in this topic?
		Pedagogy
		• How does the teacher work with the sense the children are making?
		• How does the teacher work productively with students' confusion?
		• How does the teacher attend to all students?
		• How does the teacher adjust her teaching based on the ideas she hears from students?
		Facilitating Intellectual Community
		• How does the teacher support students in showing respect for one another's ideas?
		• How does the teacher set the tone so students see each other as resources for mathematical thinking?
		• What interventions does the teacher make to ensure that students' engagement has a focus on mathematical ideas?

Resource List

Chapters and Articles

Ball, Deborah Loewenberg. (2000) Bridging Practices: Intertwining content and pedagogy in teaching and learning to teach. *Journal of Teacher Education,* 51(3), 241–47.

In this article, Ball examines how subject matter and pedagogy have been persistently divided in the conceptualization and curriculum of teacher education. The author discusses three problems that would have to be solved to bridge this gap and to prepare teachers who both know subject-matter content and can use it in making wise pedagogical choices. The problems include: identifying the content that matters for teaching, understanding how such knowledge needs to be held, and knowing what it takes to use such knowledge in practice.

Ball, Deborah Loewenberg. (1993). With an eye on the mathematical horizon: dilemmas of teaching elementary school mathematics. *The Elementary School Journal,* 93, (4).

Ball provides a powerful account of the challenges and dilemmas of developing a mathematics teaching practice that is both responsive to where children are in their mathematical understanding yet honors the long-established content of the discipline. Using examples from her own third-grade class, the author examines the kind of dilemmas that arise in three areas: the mathematical content itself, respecting children as mathematical thinkers, and creating and using community.

Ball, Deborah Loewenberg. (1992) Magical hopes: Manipulatives and the reform of math education. *American Educator, 16*(2), 14–18, 46–47.

Citing examples from her third grade mathematics classroom, the author describes problems stemming from the use of concrete objects or "manipulatives" in mathematics classrooms (e.g., fraction bars, base-10 blocks, and Popsicle sticks). The vignettes show the fallacy of assuming that students will automatically draw the conclusions that their teachers want simply by interacting with particular manipulatives.

Conference Board of the Mathematical Sciences. (2001) Recommendations for Elementary Teacher Preparation, Chapter 3 in *The Mathematical Education of Teachers.* Providence, RI: American Mathematical Society.

In this chapter, the authors describe what teachers need to understand in order to teach elementary mathematics (numbers and operations; algebra and functions; geometry and measurement; and data analysis, statistics, and probability). The chapter opens with a vignette from a third grade classroom about a teacher, who instead of taking her students through an algorithm step by step, probes their ideas in order to understand their thinking. The decisions the teacher makes and the thinking that underlies them are described, as are the understandings that the teacher needed to have to make these pedagogical decisions. How to transform a poorly prepared prospective elementary teacher into someone who can think mathematically is also addressed in this chapter.

Economopoulos, Karen. (1998) What comes next?: The mathematics of pattern in kindergarten. *Teaching Children Mathematics, 5*(4), 230–233.

This author discusses the purposes of developing facility with patterns in kindergarten. She illustrates how pattern activities, when well designed, can help children begin to develop their thinking about such complex mathematical ideas as predictability and consistency. She illustrates how such activities in the early grades help students connect with the mathematics of the later grades.

Goldenberg, E. Paul (2000) *Thinking (and talking) about technology in math classrooms.* Newton, MA: Education Development Center.

The author provides a framework for making decisions about technology use in mathematics classrooms. He points to research findings to stress that the value of technologies, from manipulatives used in the early grades to complex computer programs used in the higher grades, depends on how they are used. He suggests that dimensions such as genre of the technology, purpose of the lesson, nature of the thinking being asked of students, the role of the technology in the lesson, and prioritizing content and knowledgeable use of technology are important elements to consider.

Grant, Cathy Miles. (2000). Beyond just doing it: Making discerning decisions about using electronic graphing tools. *Learning & Leading with Technology, 27*(5) 14–17, 49.

Drawing on a project that incorporates both hand-drawn and computer-generated graphs to enrich second graders' understanding of data and what they represent, this author notes that the value of electronic graphing tools depends on teachers' recognition of the strengths and limitations of their use in the classroom.

Grant, C. M., and Lester, J. (2001) Mathematics supervision through a new lens, *Educational Leadership, 58*(5), 60–63.

Using illustrations from a *Lenses on Learning* seminar that took place in Western Massachusetts, these authors examine the kind of learning that can take place when school administrators focus on the mathematics content of the classroom. They describe the structure of the course and highlight how it helps participants to consider new ideas about mathematics, learning, and teaching, as well as their own needs for professional development within a standards-based context.

Hiebert, J., & Stigler, J. W. (2000). A proposal for improving classroom teaching: Lessons from the TIMSS video study. *Elementary School Journal, 101,* 3–20.

These authors discuss results from the TIMSS Video Study regarding U.S. teachers' view of reform. They presents the view of many teachers that they are changing the way they teach even though the core of their practice remains the same. The authors explore why this is the case and why it can be challenging to change such patterns of teaching in the United States. They conclude by offering suggestions for developing school-based, teacher-driven systems for improving teaching.

Lewis, C. & Tsuchida, I. (1998). A lesson is like a swiftly flowing river: Research lessons and the improvement of Japanese education. *American Educator,* Winter, 14–17 & 50–52.

These authors examine the structure of Japanese "research lessons," or "study lessons," which form the core of the Japanese practice of "Lesson Study." The authors describe the components of a research lesson and examine the impact that such lessons have in the improvement of Japanese education. They also consider what the key social, cultural, and political conditions are that support the lesson study system, in comparison with those of the United States.

Nelson, B. S. & Sassi, A. (2000) Shifting approaches to Supervision: The case of mathematics supervision, *Educational Administration Quarterly, 36*(3), 553–584.

These authors examine how supervisory practices have not taken account of subject-matter content but have focused primarily on pedagogical process. They address ways that administrators can better support standards-based instruction by shifting their approaches to supervision to attend to the intersection of process and content. The authors report on a study that looked at what administrators thought significant when viewing the same videotape of a fifth-grade mathematics lesson at the beginning and end of a professional development seminar on supervision. The authors conclude that administrators need to have adequate subject-matter knowledge for doing supervision and suggest several possible directions for achieving this shift.

Reitzug, Ulrich C. (1997) Images of Principal Instructional Leadership: From Super-Vision to Collaborative Inquiry. *Journal of Curriculum & Supervision.* vol. 12 no.4 (pp. 324–343.)

The author examines images of principals' instructional leadership, based on 10 supervision textbooks published between 1985 and 1995. He looks at how textbooks have portrayed principals as experts and superiors, teachers as deficient and voiceless, teaching as fixed technology, and supervision as a discrete intervention. Images of professional growth (that stress collegiality and continuous improvement) suggested by studies of successful schools differ significantly from these textbook images.

Schifter, D. (1999). Reasoning About Operations: Early algebraic thinking, grades K–6, in L. Stiff & F. Curio (Eds.) *Mathematical Reasoning, K–12: 1999 NCTM Yearbook* (pp. 62–81), Reston, VA: National Council of Teachers of Mathematics.

The author takes a close look at the development of students' operation sense in K–6 classrooms. She makes a case for the importance of organizing classrooms around students' mathematical ideas so that they can build understanding of the algebra they will encounter in later grades. The author describes a number of scenarios from elementary classrooms in which teachers are working to align their instructions with standards-based practice and which support students to articulate their mathematical ideas. She suggests that what can be learned about the development of students' operation sense from these scenarios can provide useful information about preparing young children for algebra.

Spillane, J., Halverson, R., & Diamond, J., (2001). Investigating school leadership practice: A distributed perspective. *Educational Researcher, 30*(3), 23–28.

These authors propose a framework for investigating how school leadership is "distributed" or stretched over the school's social and situational contexts. The authors ground their conceptualization of "distributed leadership" in activity theory and distributed cognition and lay out directions for a research program in which the nature of distributed leadership can be observed and analyzed.

Tracy, Saundra J. (1995). How historical concepts of supervision relate to supervisory practices today. *Clearing House. 68*(5), 320–325.

The author describes seven phases in the evolution of supervisory practice in the schools. She looks at each historical phase in relation to its purpose (assisting or assessing), focus or emphasis, the personnel typically involved, the skills needed to implement supervision, and the assumptions surrounding the process.

Books

Driscoll, Mark. (1999). *Fostering Algebraic thinking: A guide for teachers grades 6–10.* Portsmouth, NH: Heinemann.

This book is intended for teachers who want to reflect on their thinking about the teaching and learning of pre-algebra and early algebra. It builds on the move to base the teaching and learning of mathematics on explicit standards. The author presents a detailed description of his perspective on algebraic thinking and provides a framework for the use of classroom questions that foster the development of algebraic thinking in students from grades 6–12.

Elbow, Peter. (1986). *Embracing contraries: explorations in learning and teaching.* New York: Oxford University Press.

Mr. Elbow argues that what is actually most natural in teaching and learning is a rich messiness of paradox and contradiction. Consequently, he points out that we need to alter our view of how people learn and how teachers should teach and grade. In the book, he explores the learning process, the teaching process, the evaluation process, and the nature of inquiry.

Fosnot, C.T., and M. Dolk. (2001). *Young Mathematicians at Work: Constructing Number Sense, Addition, and Subtraction.* Portsmouth, NH: Heinemann

Fosnot, C.T., and M. Dolk. (2001). *Young Mathematicians at Work: Constructing Multiplication and Division.* Portsmouth, NH: Heinemann

Fosnot, C.T., and M. Dolk. (2001). *Young Mathematicians at Work: Constructing Fractions, Decimals, and Percents.* Portsmouth, NH: Heinemann

In these three books, the authors present students' approaches to solving mathematical problems, and they describe how teachers work to support and develop their students' mathematical thinking. Fosnot and Dolk lay out and explain the big ideas that comprise an understanding of the mathematical concepts addressed in each book.

Glickman, Carl D. (1992). *Supervision in Transition.* Washington, D.C.: Association for Supervision and Curriculum Development.

This collection of essays, written by leading scholars in the field of supervision, explore the connections between the current changes in school organization and governance structures and the supervisory skills and relationships that are called for by these changes. The collection provides an historical overview of the supervisory process, explorations of promising practices, and consideration of the preparation of teachers. The book acknowledges the need for supervision to entail professional inquiry and collegiality.

Glickman, C. D., Gordon, S. P., and Ross-Gordon, J. M. (1998) *Supervision of Instruction: A Developmental Approach.* Boston: Allyn and Bacon.

This textbook offers a "developmental approach" to the practice of supervision. It is built on the assumption that the aim of supervision is to help teachers develop as reflective, autonomous professionals and that supervision itself should ultimately be non-directive. The authors provide a variety of positions, ideas, and practices as well as theoretical grounding and case examples.

Sergiovanni, Thomas J. & Starratt, Robert J., (1993). *Supervision: a redefinition, fifth edition.* Boston, MA: McGraw-Hill.

The authors propose a reconceptualization of the supervisory role and its place within the school community. In this edition, they place professional and moral authority rather than bureaucratic authority as the central force behind what teachers should do and how supervision should be done. They replace the metaphor of organization with that of community to describe the nature of schools and schooling. While still emphasizing the importance of traditional supervisory skills and practices, the authors stress the moral relationship that needs to be established between supervisor and teachers for these skills to be exercised wisely.

Stigler, J. W., & Hiebert, J. (1999). *The teaching gap: Best ideas from the world's teachers for improving education in the classroom.* New York, NY: Summit Books.

Drawing on the conclusions of the Third International Mathematics and Science Study (TIMSS), Mr. Stigler and Hiebert offer an action plan for improving education in the United States. The TIMSS study used video to observe a large number of classroom lessons in the U.S., Japan, and Germany. In the book, the authors use data from the study, especially from eighth-grade mathematics classrooms, to examine why the quality of teaching in the U.S. lags behind that of our peers in other countries. Drawing particularly from the Japanese practices, including Japanese "lesson study," the authors argue that U.S. schools can be restructured as places where teachers can engage in career-long learning and classrooms can become laboratories for developing new teaching-centered ideas.

Curricula

Connected Mathematics Project (CMP) is a mathematics curriculum for grades 6–8. It was developed at Michigan State University with funding from the National Science Foundation, and it is published by Prentice Hall. The materials provide opportunities for students to investigate mathematics concepts that come from everyday situations.

Investigations in Number, Data, and Space is a K–5 mathematics curriculum developed at TERC in Cambridge MA with funding from the National Science Foundation. Investigations provides students with in-depth experiences in number, data, geometry, and the mathematics of change. It engages students in the exploration of major mathematical ideas and supports them to develop their own approaches to solving problems based on their knowledge and understanding of mathematical relationships. The curriculum also offers support for teachers in both mathematics content and pedagogy.

Videotapes

Mathematical Inquiry through Video: Tools for professional growth. A package of videos and teacher development materials developed by BBNT Solutions LLC.

This video package consists of ten video cases of middle school mathematics classrooms. They depict teachers who are working to change their teaching practice according to the NCTM standards and highlight the challenges, ideas, and issues these and other teachers face in this process. Each video is accompanied by a facilitator's guide which includes a description of the video; background information on the school, teacher, and classroom in the video; suggested workshop design features with video-related mathematical and pedagogical issues and activities; and a complete transcript. For more information on this series, you may contact BBN at 10 Moulton Street, Cambridge, MA. Or you may wish to speak with one of the authors of the materials, Fadia Harik, at f.harik@comcast.net or fadia.harik@umb.edu.

Mathematics: Assessing Understanding. A Series of Videotapes for Staff Development, featuring Marilyn Burns. While Plains, NY. Cuisenaire Company of America, Inc.

This series consists of three videotapes and accompanying teacher's discussion guide, showing a collection of individual assessments of mathematical understanding with students ages 7 through 12. All of the assessments address students' ability to estimate, reason numerically, and compute in problem-solving situations. The one-on-one interviews model for teachers the kinds of questions that are useful for gaining insights into how students are thinking and what they understand. Mathematical topics included are: number sense and the place value structure of our number system; estimation, numerical reasoning, and computation with whole numbers; and fractions.

Relearning to Teach Arithmetic. A series of videotapes for staff development, funded by the NSF and developed at TERC, Cambridge, MA. Published by Dale Seymour Publications, White Plains, NY: an imprint of Addison Wesley, Longman, Inc.

This video series provides teachers with a structured opportunity to explore how children develop facility with the four operations (addition, subtraction, multiplication, and division) and how teachers can foster the development of this facility. It consists of two packages, each containing 3 videos. The first package focuses on addition and subtraction and the second focuses on multiplication and division. Each package also includes a study guide that outlines six professional development sessions. During each session, teachers view and discuss segments of the tapes and work on related mathematics problems. Each package is adaptable to a variety of settings such as after-school sessions, release-day professional development sessions, or summer staff development experiences. The materials may also be integrated into a longer course or seminar on the teaching and learning of elementary mathematics.

Talking Mathematics. A professional development resource package funded by the NSF and developed at TERC, Cambridge, MA. Published by Heinemann, Portsmouth, NH.

This package includes a videotape program, a resource guide for staff developers and university instructors, and a book for teachers who are interested in supporting talk and mathematical inquiry in their classrooms. The goal of the package is to provide teachers and staff developers with resources that can help them cultivate good mathematical discourse. The video program consists of an introductory videotape, four twenty-minute videotapes on aspects of children's talk, six short classroom episodes, and a twenty-minute summary of a Talking Mathematics teacher seminar. The package can be adapted to a variety of professional development settings.

Teaching Math Video Libraries. A series of videotapes for staff development, funded by the Annenberg Foundation/Corporation for Public Broadcasting and produced by WGBH (1995). S. Burlington, VT: The Annenberg/CPB Math and Science Collection, (800) 965-7373.

This video series provides visual examples of standards-based teaching and learning. Four Teaching Math "libraries" are available: K–4, 5–8, 9–12, and a K–12 assessment library. The grade level libraries each consist of a set of content standard videos, a set of process standard videos, and guidebooks. The assessment library provides case study videos that examine assessment issues in two different classes and a sequence of vignettes that show a variety of assessment techniques from several classes. This collection of video libraries provides grounded images of what classrooms may look like when teachers are developing their teaching in accordance with NCTM standards.

Notes

Notes

NOTES

Notes

Notes

Notes

Notes

Notes

Notes

Notes